CIRCUIT MODELING:

EXERCISES AND SOFTWARE

Third Edition

with

***BREADBOARD*™**
A Circuit Analysis Computer Program
with Simulation Graphics

F.A. Ciccarelli

Niagara College of Applied Arts and Technology

and

Bayfield Engineering Software

Prentice Hall
Education, Career & Technology
Englewood Cliffs, New Jersey   Columbus, Ohio

Library of Congress Cataloging-in-Publication Data

Ciccarelli, F. A.
   Circuit modeling : exercises and software : with Breadboard, a circuit analysis/computer
   program with simulation graphics / F. A. Ciccarelli.—3rd ed.

      p.   cm.
      ISBN 0-02-322473-8 (pbk.)
      1. Breadboard. 2. Electric circuits—Computer simulation. 3. Electronic circuits—Computer
   simulation. I. Title
TK454.C59 1995
621.319'21'078–dc20   94-9549
                                                                   CIP

Editor-in-Chief: *Greg Burnell*
Acquisitions Editor: *Dave Garza*
Developmental Editor: *Carol Hinklin Robison*
Production Editor: *Patricia A. Skidmore*
Production Manager: *N. Corinne Folino*

Prentice-Hall International (UK) Limited, *London*
Prentice-Hall of Australia Pty. Limited, *Sydney*
Prentice-Hall Canada Inc., *Toronto*
Prentice-Hall Hispanoamericana, S.A., *Mexico*
Prentice-Hall of India Private Limited, *New Delhi*
Prentice-Hall of Japan, Inc., *Tokyo*
Simon & Schuster Asia Pte. Ltd., *Singapore*
Editora Prentice-Hall do Brasil, Ltda., *Rio de Janeiro*

DEDICATION

To my grandchildren,
Kristen, Brett, and Dylan,
a joy in my life and a vision of the future.

The computer program *Breadboard* and the exercises of this manual were written to enhance the study of electrical circuit analysis. The exercises provide an orderly sequence of topics to reinforce the concepts taught in the lecture room. The goal of circuit analysis is to obtain an understanding of the conditions that exist in a circuit and the parameters that affect those conditions. Computer analysis reduces the mathematical requirements of effecting a solution but does not replace the need to study the solution results in order to understand the relationships. Once the fundamentals have been mastered computer analysis provides the means to extend the topics of study to circuits that would otherwise require extensive mathematical computations. "What if" type questions can easily be resolved with the use of computer analysis.

The exercises of this third edition of exercises to accompany the software program *Breadboard* have been completely reworked and supplemented. Each exercise is designed to reinforce the key concepts related to the objectives of the topic by requiring the student to respond to questions, fill in tabular results, or obtain and analyze graphs. The added exercises extend the study of circuit analysis to a more advanced level of circuit analysis than previous editions.

The *Breadboard* circuit analysis program has been colorized for improved visual effect and increased capabilities have been added. The number of points plotted in the graph option has been increased from 30 to 120 points to provide higher resolution. The choice of independent variable has been extended to voltage sources and current sources as well as resistance for both the AC and DC analysis options. Bode plots as well as regular graphs as a function of frequency are now available in the AC option. Other improvements include a laser printer driver, schematic printout in landscape or portrait form, display of the file names of circuits saved on disk when the load option is chosen, and a graphic representation for choosing the graph options. In the AC module a calculate option has been added to convert reactance and impedance to R, L, and C.

ACKNOWLEDGMENTS

The support received from the staff at Prentice Hall Career & Technology in Columbus in developing this manual is especially appreciated. In particular Carol Robison, Patricia Skidmore, and Dave Garza have my sincere thanks for their direction and confidence.

CONTENTS

Familiarization with the BREADBOARD Program

OBJECTIVE

To learn to use the *Breadboard* circuit-analysis program.

BACKGROUND

The *Breadboard* circuit-analysis program is capable of providing steady-state solutions for analog DC and AC circuits. The program enables the user to draw the circuit schematic directly on the screen of the computer monitor and solve for the circuit parameters of current, voltage across each element, voltage as measured to the ground (common) reference point, and power in the resistive elements. In AC circuits, the phase angle of the parameters is also given, as well as the reactive volt-amperes. The passive components, R, L, and C are provided, as well as constant and controlled (dependent) voltage and current sources. The program offers graphing capabilities and printing of the schematic, graph, and solution results.

PROCEDURE

The appendix of this manual contains a complete reference for the program and should be read prior to this familiarization exercise. The program is very easy to use and prompts the user on the second line of the screen at each step. The prompt line should be referred to as required until the user has mastered the program.

STARTING the *BREADBOARD* PROGRAM

From a Floppy Drive:

Turn the computer on.
Insert the *Breadboard* program disk in drive A (or B).
Change the default drive by typing A: (or B:) and pressing <ENTER>.
To start the program type BB and press the <ENTER> key.

Computer with hard drive:

Turn on the computer and obtain the C:>\ prompt.
Create a suitable sub directory (e.g. BB) by typing MD BB and <ENTER>.
Change the default directory to C:>\BB by typing CD BB and <ENTER>.
Copy all the files on the *Breadboard* disk to the directory (type COPY A:*.*).
To start the program type BB and press the <ENTER> key.

The *Breadboard* logo will appear on the screen, followed by the Main Menu, which provides a choice of DC or AC circuit analysis or of exiting the program back to the system.

> For this practice session dc analysis will be used.
> Press the <ENTER> key to select the dc option.

On the top line of the screen will appear the following menu options used to draw and solve a dc circuit. Menu options are selected by using the cursor-control keys on the keyboard to move the blinking selection box to the option desired and pressing <ENTER>.

> | DRAW | SOLVE   NEW   SAVE   LOAD   PRINT   GRAPH   EXIT

Also on the screen is a grid of dots that outlines the limits of the breadboard. The rectangular window at the bottom of the screen is the Information Window, which will display component values or solution results.

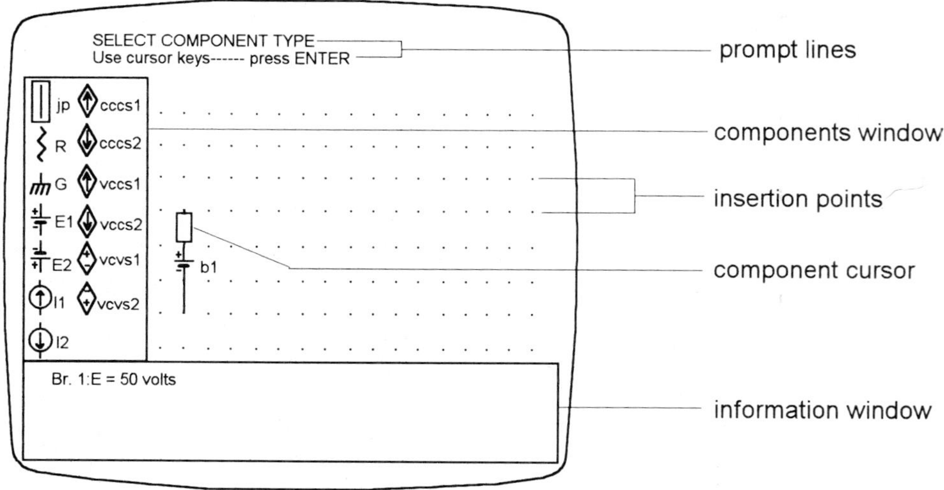

Figure 1.1

The circuit to be analyzed is to be drawn on the screen. Select the DRAW option by pressing the <ENTER> key. A small rectangle, called a **component cursor**, will appear in the lower left part of the screen. This component cursor determines where on the breadboard a component is to be placed. The component cursor can be moved around on the breadboard by using the LEFT, RIGHT, UP, and DOWN arrow keys on the keyboard. Do it now. Move the cursor all over the screen and then return it to its initial position.

The process of drawing the circuit consists of the following steps:

1. Position the cursor to the place where a component is to be drawn and press <ENTER>.

2. A component selection window will appear on the left of the screen (Figure 1.1).

3. Select the desired component by using the cursor control (arrow) keys on the keyboard.

4. Press <ENTER> to have that component drawn in the position of the component cursor.

The circuit to be analyzed for this practice session is the simple series circuit shown in Figure 1.2. The initial position of the component cursor is a convenient place to start the drawing on the breadboard. The first component needed is a connecting wire labeled jp (jumper) in the component window. Press <ENTER>. The component window appears with the flashing selection box located around the symbol for a connecting wire. Press <ENTER>. The wire is drawn in the position where the component cursor was located, and the cursor has automatically moved upward to the next position.

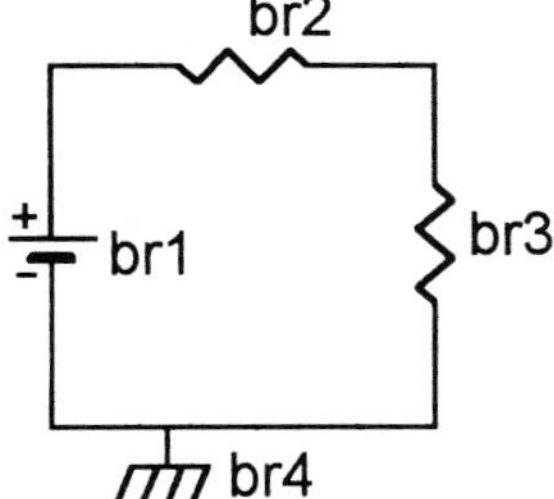

Figure 1.2

Net List for the Circuit

Br. 1: E = 60 V
Br. 2: R = 10 Ω
Br. 3: R = 10 Ω
Br. 4: voltage reference

A voltage source is to be placed in this new position. Press <ENTER> to obtain the component window. Use the cursor keys to select the voltage source symbol from the window and press <ENTER>. The voltage source has now been drawn in the last cursor position, and the cursor has moved upward again. (See Figure 1.1.) The prompt line at the top of the screen now requests a value to be entered for the magnitude of the voltage source. Type in the value 60 and press <ENTER>. Do not type in the units V or Volts; only the numeric value is required. Note that the Information Window now displays the value and that this component has been designated as branch 1 (br1) on the screen.

Continue the procedure of drawing the circuit until it is completed (Figure 1.2). It is best to draw the components in a sequence that follows the direction of the expected current flow. Each circuit **must** have a ground reference point, as shown in Figure 1.2 as branch 4 (br4). The ground reference symbol is available only in the vertical components window since ground or chassis common symbols are drawn vertically in schematics.When the drawing is complete return to the menu by pressing the <ESC> key.

The circuit is ready to be solved. Select the SOLVE option by pressing the right-cursor key to move the blinking selection box over the SOLVE option, and press <ENTER>. The program will solve the circuit and the solution results will appear in the Information Window. For this circuit the solution for the current will be 3.00 amperes in each branch. To see the voltages and power for each component scroll the solutions in the information window. Use the PgUp and PgDn keys on the keyboard. To return to the menu, press the <ESC> key.

EDITING the CIRCUIT

Changing the value of a component or deleting components on a schematic on the screen is easily done using the editing features of the program. To change the value of a component the component must be erased and a new component drawn in its place. Choose the DRAW option from the menu and move the component cursor over the component to be changed. Pressing the <ENTER> key will erase that component, and the Information Window will indicate that the branch has been deleted. To insert a new component, press <ENTER> again to obtain the Component Selection Window, make the selection as before, and the new component will be drawn in place of the cursor. Then enter the value of the component as requested by the prompt line.

Following the above procedure change the value of the resistor in branch 3 to 20 ohms and obtain the solution to the new circuit. The solution results will show the current has changed to 2.00 amperes. Press <ESC> to return to the menu.

NEW, SAVE and LOAD

To save the circuit configuration onto the disk choose the SAVE option. The prompt line will request a name for this circuit. Type DCEXER1 and press <ENTER>. The circuit is now saved on the disk and the program returns to the menu.

To erase a circuit from the screen and from memory select the NEW option and press <ENTER>. The menu will reappear and the circuit will have been erased from the screen. The user may now choose to draw a new circuit or recall a circuit from memory.

Choose the LOAD option from the menu. The circuits stored on the disk will be listed on the screen. As only DCEXER1.CKT will be on the disk, this will be the only one listed. When more than one is listed the user selects the circuit using the cursor control keys. Press <ENTER> and the circuit will be drawn on the screen ready for solution.

PRINTING the SOLUTION RESULTS

```
Schematic
Net List
Br. Currents
Br. Power
Node Voltage
Br. Voltage
** PRINT **
<ESC> to menu
```

Figure 1.3

The PRINT option allows the printing of the solution results and the circuit schematic on the attached printer. The program will first ask for the name of the circuit, the student's name, and the date, to create a header for the printout. A list of print options will then be displayed which allows the user to select those solution results to be printed out. Selection is made in the usual way by positioning the flashing box over the desired item and pressing <ENTER>. A selection may be unselected by repositioning the flashing box over that item and pressing <ENTER>. When all selections have been made move the flashing box down to the word *print* and press <ENTER>. The selected solution results will be printed and the menu will then reappear.

The GRAPH OPTION

For this exercise a graph of the power in the resistor of branch 3 is to be obtained as the value of the resistor in branch 3 is varied from 20 to 80 ohms.

To facilitate selecting the X-Y parameters to be plotted, a representation of a graph is displayed in the Information Window when the GRAPH option is selected. The X-axis variable (independent variable) is to be selected first and a flashing box is displayed around the word *resistance* below the X-axis. Use the

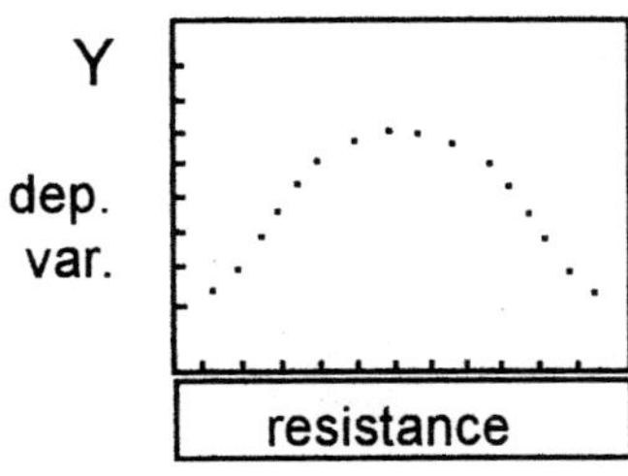

Figure 1.4

cursor keys to toggle through the other options, source voltage and source current, and then return to *resistance* as this is to be the X-variable. Press <ENTER>. The prompt line will request the branch number of the resistor to be varied. Enter 3. The prompt line will then request the range of variation of the value of this resistor. For the lower limit (the beginning value) respond with 20 and for the upper limit (ending value) enter the value 80.

The flashing selection box will then be located on the Y-axis parameter (dependent variable) to be plotted. Again toggle through the parameters of branch current, branch voltage, node voltage, and branch power. Branch power is the Y-parameter to be plotted. Press <ENTER>. The prompt line will request the branch number whose power is to be plotted. Enter 3.

The program will automatically scale the axes, solve the circuit for 120 values of the resistance in branch 3, and plot a full screen graph. The prompt line asks if the user wishes to print a copy of the graph on the printer. Respond by typing N for no. The next prompt will ask whether a printout of the X and Y values is desired. Type N for no. To return to the menu type Y for yes.

To leave the program and return to DOS select EXIT from the main menu. The choice of DC, AC, or EXIT will be displayed. Choose EXIT.

Ohm's Law

OBJECTIVE

To apply Ohm's law and to show the linear relationship of the voltage across and the current through a resistor in a DC circuit.

THEORY

Ohm's law establishes the fundamental relationship between the voltage across and current flowing through a resistor in an electrical circuit. It is basic to all further analysis of electric circuits and the student must become proficient in its use in the three forms given below where R is the magnitude of the resistor, V is the voltage across, and I is the current through the resistor.

$$V = IR \qquad\qquad I = V/R \qquad\qquad R = V/I$$

Mathematically, Ohm's law is described as a linear relationship because for a given value of resistance, increasing the voltage will result in a proportional increase in current. If a graph is plotted of the current through the resistor versus the value of the voltage source a straight line will be obtained. Note that if the range of variation of the voltage source is large, the graph will appear as a curve, but for any point on the graph Ohm's law will hold true.

PROCEDURE Part 1

This procedure shows the effect of changing the resistance on the current flowing in a circuit with the voltage of the source held constant. Using the *Breadboard* program, draw the circuit shown in Figure 2.1 with the component values shown in the net list for the circuit. Save the circuit as DCEXER2.

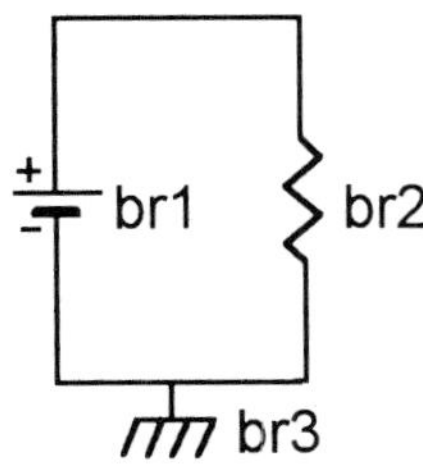

Net List for the Circuit

Br. 1: E = 240 V
Br. 2: R = 10 Ω
Br. 3: voltage reference

Figure 2.1

Solve the circuit and enter the value of current in the table below. Prove the Ohm's law relationship for voltage and current by calculating the voltage E and entering its value in the table.

Table 2.1

$E = V = 240$ volts		
Resistance	Current I	$V = E = IR$
10 Ω		
200 Ω		
4 kΩ		
50 kΩ		
120 kΩ		
6 MΩ		

Edit the circuit, changing the value of the resistor to each of the remaining values in Table 2.1 . Solve the circuit for each value and complete the table.

Study the tabulated results and complete the following statements.

As the resistance was increased the current ________________.
When the resistance increased from 10 to 200 ohms the current ____________
by a factor of ________________.

If the resistance is held constant and the source voltage is varied the current will change in value in accordance with Ohm's law. Edit the circuit of Figure 2.1 to change the resistance value to 200 ohms and the voltage of the source to 20 volts. Obtain a solution for the circuit for each value of source voltage shown in Table 2.2. Complete the table.

Table 2.2

Resistance = 200 Ω		
Voltage $E = V$	Current I	$R = V/I$
20 V		
40 V		
100 V		
200 V		
400 V		
2000 V		

Study the tabulated results and complete the following statements.

As the applied voltage increased the current ________________.
When the source voltage increased from 100 to 200 volts the current ____________ by a factor of __________.

PROCEDURE Part 2

The voltage across and the current through every resistor in a circuit is related by Ohm's law. Draw and solve the circuit shown in Figure 2.2 below using the component values in the net list.

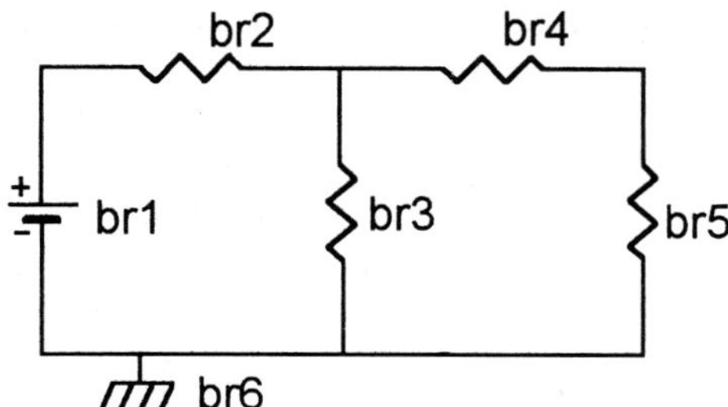

Net List for the Circuit

Br. 1: E = 100 V
Br. 2: R = 10 Ω
Br. 3: R = 20 Ω
Br. 4: R = 5 Ω
Br. 5: R = 15 Ω
Br. 6: voltage reference

Figure 2.2

From the solution results fill in the values for the current through and the voltage across (branch voltage) each resistor in the boxes of Figure 2.3 below. Complete the table by calculating the voltage, current, and resistance for each component.

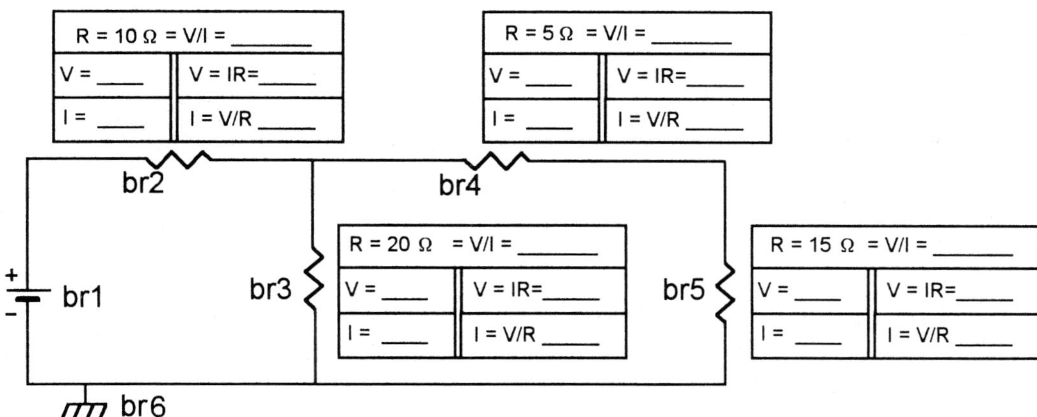

Figure 2.3

PROCEDURE Part 3

A graph of the current through a resistor as the source voltage is varied will show the linear nature of Ohm's law. Draw the circuit of Figure 2.1 and then choose the GRAPH option. Obtain a graph of the current through the resistor as the source voltage varies over a range of values from 20 to 80 volts.

Have the program print out a tabular listing of the values used to plot the graph. Confirm that the relationship $E = IR$ is true for three (3) of the points on the graph using the tabulated values.

Table 2.3

E	I	R	$E = IR$

PROCEDURE Part 4

A constant current source is the counterpart of a voltage source. The current source establishes the magnitude of the current flowing in a circuit regardless of the value of the resistance. For a given resistor the voltage across the resistor will be determined by Ohm's law.

Using the *Breadboard* program draw the circuit shown in Figure 2.4 for the values given in the net list and solve.

Figure 2.4

Net List for the Circuit

Br. 1: I = 4.6 mA
Br. 2: R = 1800 Ω
Br. 3: voltage reference

Complete the following statements.

The voltage across the resistor is _______________ V.

By Ohm's law the voltage is $V = IR =$ _______________ V.

FURTHER ANALYSIS

Solve several problems in your text in the section on Ohm's law and verify your solutions using *Breadboard*.

Voltage Sources in Series

OBJECTIVE

To confirm that in a circuit containing several voltage sources connected in series, the net voltage acting in the circuit is the algebraic sum of the individual source voltages.

THEORY

Voltage sources are connected in series if the same current flows through them. The polarity of the voltage sources connected in series determines whether their voltages are series aiding or series opposing. If the polarities are the same for a chosen direction of current flow, the sources are series aiding and their magnitudes are added to obtain the net voltage acting in the circuit. If the voltage sources are series opposing their values are subtracted.

PROCEDURE Part 1

For the circuit shown in Figures 3.1, use *Breadboard* to solve for the current in the circuit and the voltage at node 2.

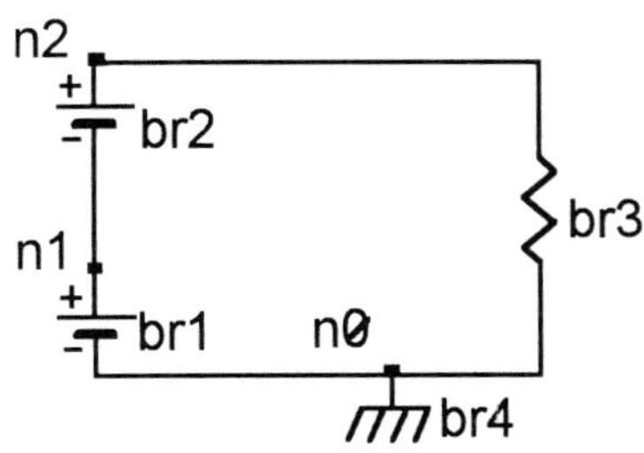

Net List for the Circuit

Br. 1: E = 12.6 V
Br. 2: E = 9.3 V
Br. 3: R = 1000 Ω
Br. 4: voltage reference

Figure 3.1

Mark on the circuit diagram the direction of the current as given by *Breadboard*. The direction is represented as *node from* → *node to* in the square brackets given in the solution for the branch current in the Information Window.

Current in the circuit is ______________ mA.
Voltage at node 2 is _________ V.
Algebraic sum of the voltage sources is _________ V.
The net voltage may also be verified using Ohm's law as $V = IR =$ _________ V.

PROCEDURE Part 2

For the circuit shown in Figures 3.2, use *Breadboard* to solve for the current in the circuit and the voltage at node 2.

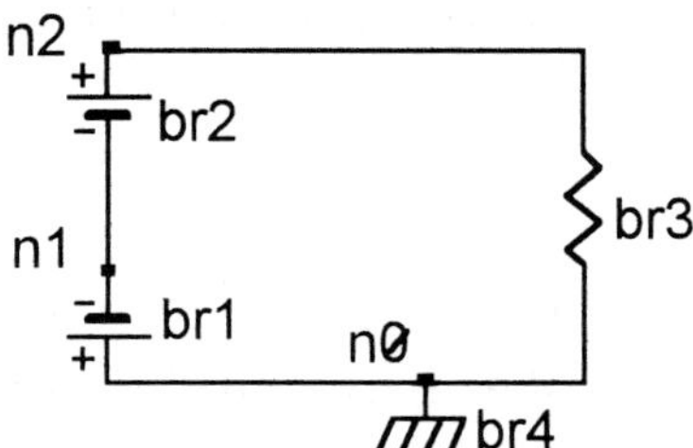

Figure 3.2

Net List for the Circuit

Br. 1: E = 12.6 V
Br. 2: E = 6.2 V
Br. 3: R = 4 Ω
Br. 4: voltage reference

Mark on the circuit diagram the direction of the current as given by *Breadboard*. The direction is represented as *node from* → *node to* in the square brackets given in the solution for the branch current in the Information Window.

Current in the circuit is ______________ A.
Voltage at node 2 is _________ V.
Algebraic sum of the voltage sources is _________ V.
The net voltage may also be verified using Ohm's law as $V = IR =$ _________ V.

PROCEDURE Part 3

For the circuit shown in Figure 3.3, use *Breadboard* to solve for the current in the circuit and for the voltage across the resistor. On the circuit diagram mark the current direction and the voltage across the resistor.

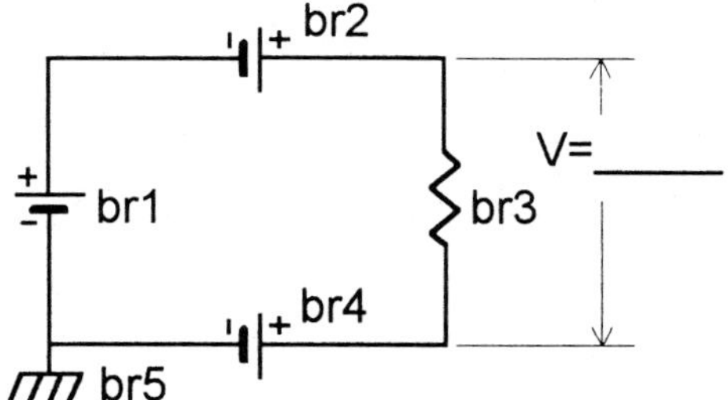

Figure 3.3

Net List for the Circuit

Br. 1: E = 15 V
Br. 2: E = 4.5 V
Br. 3: R = 40 Ω
Br. 4: E = 7.5 V
Br. 5: voltage reference

Current in the circuit is ______________ mA.
Calculate the algebraic sum of the source voltages. $E_{sum} =$ _________ V
Calculate the voltage across the resistor. $V = IR =$ _________ V

FURTHER ANALYSIS

Solve two problems in your textbook on voltage sources in series and verify your solutions using *Breadboard*.

Series Circuits
Kirchhoff's Voltage Law

OBJECTIVE

To investigate the characteristics of a DC series circuit and confirm that the total resistance of the circuit is the sum of all the resistances, and to reinforce the concept of Kirchhoff's voltage law.

THEORY

By definition, any two components are connected in **series** if the current flowing through them is common to both.

Since the current in a series circuit must flow through all the resistors connected in series, the total opposition to current flow is the numeric sum of all the resistances.

$$R_T = R_1 + R_2 + R_3 + \dots R_N$$

Kirchhoff's voltage law states that around any closed loop of a circuit, the sum of the voltage drops must equal the algebraic sum of the voltages of the sources.

$$E_T = V_1 + V_2 + V_3 + \dots V_N$$

By Ohm's law, the current that flows in the circuit is $I = E/R_T$ where E is the net voltage acting in the circuit.

The relative position of the components in the series circuit does not affect the magnitude of the current that will flow. The current magnitude will be determined by the net voltage acting in the circuit and the total resistance. A change in value of any one of the components will change the magnitude of the current.

PROCEDURE Part 1

Using the *Breadboard* circuit-analysis program, draw the DC series circuit shown in Figure 4.1 with the component values shown in the net list. Save the circuit on disk as DCEXER4.CKT.

Select the SOLVE option and obtain the solution for the circuit. From the solution results, mark the voltage across each component on the circuit diagram. Complete Table 4.1.

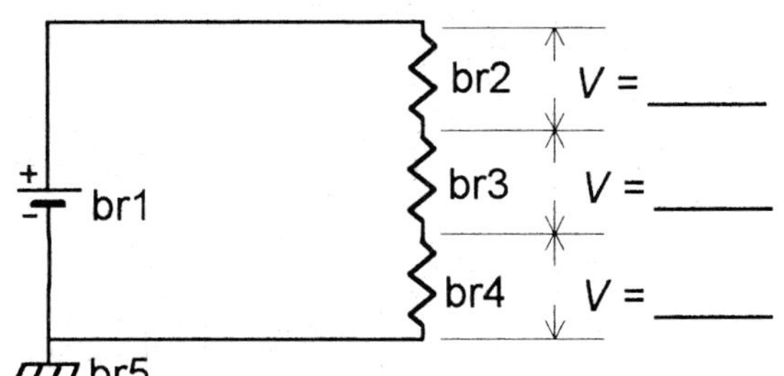

Net List for the Circuit

Br. 1: E = 45 V
Br. 2: R = 800 Ω
Br. 3: R = 500 Ω
Br. 4: R = 200 Ω
Br. 5: voltage reference

Figure 4.1

Table 4.1

R	I	$V = IR$
800 Ω		
500 Ω		
200 Ω		
$R_T =$	$I = E/R_T =$	$E =$

Confirm the definition of a series circuit.

The current through each component is _______________ mA.

Confirm Ohm's law

The calculated value of $I = E/R_T =$ _____________ mA.

Confirm Kirchhoff's Voltage Law

$E =$ sum of the voltage drops = _________________ V.

Does the largest resistor have the largest voltage drop? _________

Does the smallest resistor have the smallest voltage drop? _________

PROCEDURE Part 2

Change the value of the resistor in branch 3 to 2000 ohms. Solve the circuit and complete Table 4.2

Table 4.2

R	I	$V = IR$
800 Ω		
2000 Ω		
200 Ω		
$R_T =$	$I = E/R_T =$	$E =$

Confirm the definition of a series circuit.

The current through each component is _____________ mA.

Confirm Ohm's law

The calculated value of $I = E/R_T =$ ___________ mA.

Confirm Kirchhoff's Voltage Law

$E =$ sum of the voltage drops = _________________ V.

Does the largest resistor have the largest voltage drop? _________

Does the smallest resistor have the smallest voltage drop? _________

PROCEDURE Part 3

Use *Breadboard* to solve for the current through and the voltage across (branch voltage) each component in the circuits shown in Figures 4.2 and 4.3. Mark the values on the circuit diagram showing the direction of the current and the polarities of the voltages. Confirm Kirchhoff's voltage law and Ohm's law for each of the series circuits.

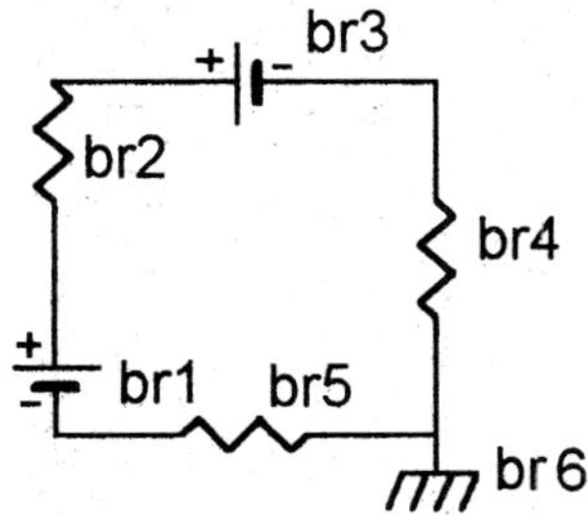

Figure 4.2

Net List for the Circuit

Br. 1: E = 50 V
Br. 2: R = 2 kΩ
Br. 3: E = 30 V
Br. 4: R = 3 kΩ
Br. 5: R = 5 kΩ
Br. 6: voltage reference

Confirm the definition of a series circuit.

The current through each component is ______________ mA.

Confirm Kirchhoff's voltage law.

Sum of the voltage drops = __________ V.

Algebraic sum of the source voltages = __________ V.

Confirm Ohm's law.

Calculated current in the circuit is $I = E_T/R_T =$ ____________ mA.

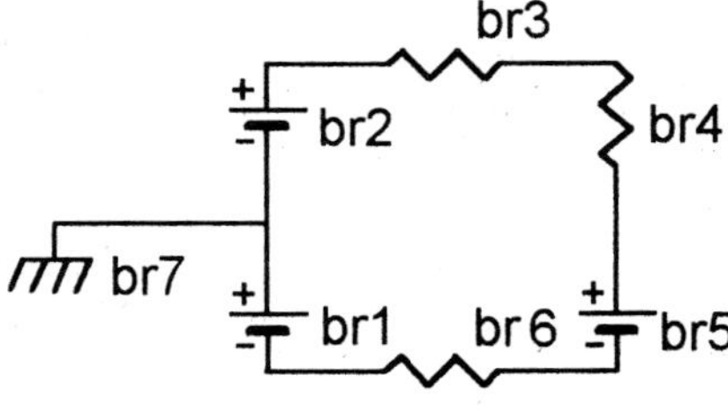

Figure 4.3

Net List for the Circuit

Br. 1: E = 250 V
Br. 2: E = 100 V
Br. 3: R = 20 Ω
Br. 4: R = 30 Ω
Br. 5: E = 150 V
Br. 6: R = 50 Ω
Br. 7: voltage reference

Confirm the definition of a series circuit.

The current through each component is ____________ A.

Confirm Kirchhoff's voltage law.

Sum of the voltage drops = __________ V.

Algebraic sum of the source voltages = __________ V.

Confirm Ohm's law.

Calculated current in the circuit is $I = E_T/R_T =$ __________ A.

FURTHER ANALYSIS

Choose two problems in your textbook from the section on Kirchhoff's voltage law. Use *Breadboard* to solve the circuits and confirm that the sum of the voltage rises around any closed loop equals the sum of the voltage drops.

Parallel Circuits
Kirchhoff's Current Law

OBJECTIVE

To study the characteristics of DC parallel circuits and confirm Kirchhoff's current law. The calculation of the total resistance of a parallel circuit by the reciprocal rule for resistors in parallel will be verified.

THEORY

Components are connected in parallel if the voltage across them is common. By Kirchhoff's current law, the sum of the currents entering a common node of a parallel circuit must equal the sum of the currents leaving the node.

The total resistance of a parallel circuit is obtained by the reciprocal rule

$$1/R_T = 1/R_1 + 1/R_2 + 1/R_3 + \dots 1/R_N$$

or by the formula for a single-source parallel circuit

$$R_T = E_T / I_T$$

The current divider rule is given by $I_X R_X = I_T R_T$

PROCEDURE Part 1

Using *Breadboard*, draw and solve the circuit shown in Figure 5.1 for the values of the components given in the net list.

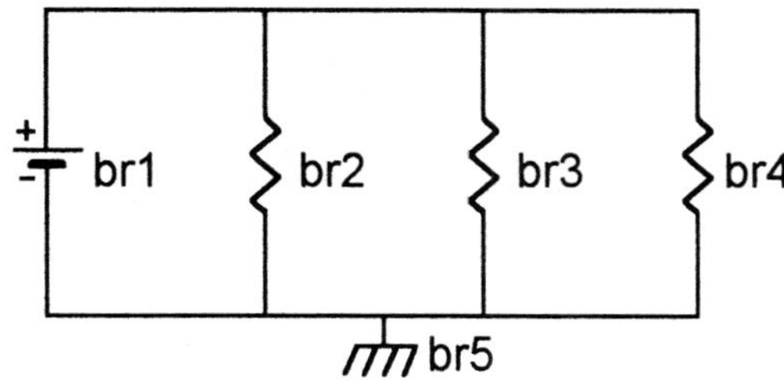

Net List for the Circuit

Br. 1: E = 48 V
Br. 2: R = 6 kΩ
Br. 3: R = 4 kΩ
Br. 4: R = 2.4 kΩ
Br. 5: voltage reference

Figure 5.1

Record on the circuit diagram the current from the source and the current in each branch, indicating the current direction as well as the magnitude.

Current from the source = I_T __________ mA

Sum of the branch currents = __________ mA

Calculate the total resistance of the circuit using the reciprocal rule for resistors in parallel.

$$R_T = \text{\_\_\_\_\_\_\_\_} \ \Omega$$

From the solution results calculate the total resistance $R_T = E_T/I_T =$ ______ Ω

Use the current divider rule to calculate the current through the resistor of branches 2 and 3.

Calculated value of $I_2 =$ ________ mA $I_3 =$ ________ mA

Solved values of $I_2 =$ ________ mA $I_3 =$ ________ mA

PROCEDURE Part 3

Draw and solve the circuit shown in Figure 5.2. Record the value of currents and their direction for each branch on the circuit diagram .

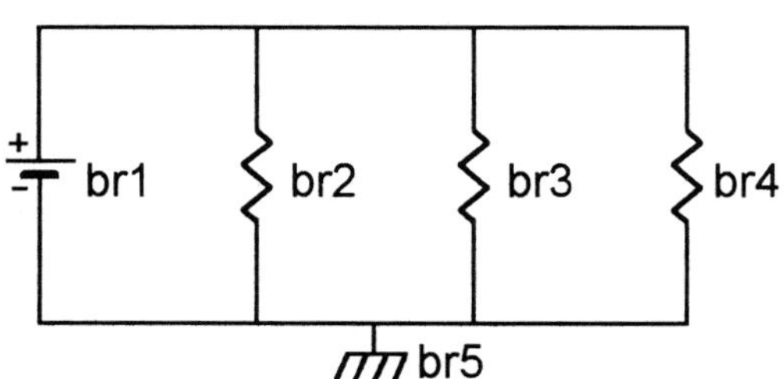

Net List for the Circuit

Br. 1: E = 36 V
Br. 2: R = 90 Ω
Br. 3: R = 1.2 kΩ
Br. 4: R = 1.8 MΩ
Br. 5: voltage reference

Figure 5.2

The total source current (branch 1) = __________ mA.

The sum of the branch currents = __________ mA.

From the solution results calculate the total resistance $R_T = E_T/I_T =$ ______ Ω

Does the lowest-value resistor pass the highest current? (yes or no) ________

Is the total resistance lower than the smallest resistor? (yes or no) _______

Comment on the relative magnitudes of the branch currents by comparing them to the source current.

Voltage Dividers
Unloaded

DC Exercise

OBJECTIVE

To investigate the concept of unloaded voltage dividers in a DC series circuit.

THEORY

The voltage across a resistor in a single source series circuit is obtained by Ohm's law as $V_X = IR_X$ and the current through the resistor is determined by the relationship $I = E/R_T$. Substituting for the current in the first equation gives the following equation.

$$V_X = \frac{E}{R_T} R_X \quad \textbf{or} \quad V_X = \frac{R_X}{R_T} E$$

The second equation gives the voltage divider rule expressed as the voltage across any resistor in a series circuit is the ratio of the value of the resistor to the total resistance of the circuit times the applied voltage.

PROCEDURE Part 1

To obtain several values of DC voltage from a single source, voltage dropping resistors are added in series with the source as shown in Figure 6.1. The voltage at each node can be calculated as E times the ratio of the resistance across that part of the circuit to the total resistance of the circuit.

Draw and solve the circuit shown in Figure 6.1 using *Breadboard*. From the solution results, record the voltage at each node.

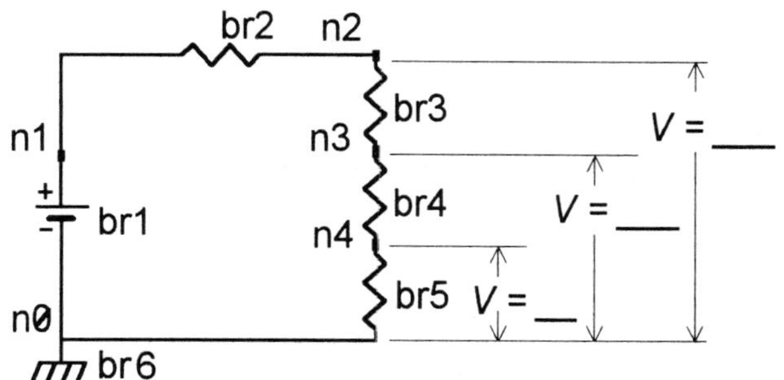

Figure 6.1

Net List for the Circuit

Br. 1: E = 24 V
Br. 2: R = 400 Ω
Br. 3: R = 500 Ω
Br. 4: R = 500 Ω
Br. 5: R = 1000 Ω
Br. 6: voltage reference

Use the voltage divider rule to calculate the following voltages.

The voltage at node 2 = $(R_3 + R_{4+}R_5)/R_T \cdot E$ ________ V.

The voltage at node 3 = $(R_4 + R_5)/R_T \cdot E$ _______ V.

The voltage at node 4 = $(R_4/R_T) \cdot E$ _________ V.

The voltage across the resistor in branch 2 = _______ V.

Compare your answers to the values obtained with *Breadboard*.

PROCEDURE Part 2

Using the *Breadboard* program draw and solve the circuit shown in Figure 6.2.

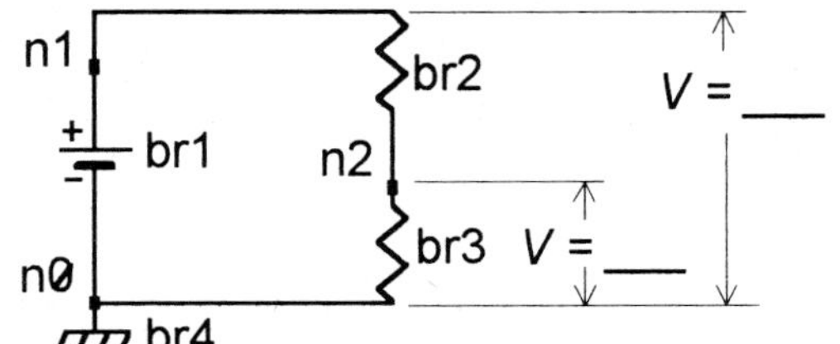

Figure 6.2

The voltage at node 2 = ________ V.

The voltage at node 2 is ______ % of the source voltage.

Change the source voltage to 50 volts and solve the circuit.

The voltage at node 2 = ________ V.

The voltage at node 2 is ______ % of the source voltage.

FURTHER ANALYSIS

Choose a voltage divider problem from your text. Design the circuit for the required resistors and then use *Breadboard* to solve for the node voltages to prove your design.

Voltage Differences

OBJECTIVE

To emphasize the concept of voltage difference between two points in a DC circuit.

THEORY

The word *voltage* implies a difference of potential between two points in a circuit. To designate the voltage difference between two points, a double subscript notation is used. The voltage difference between two points labeled **a** and **b** would be designated as V_{ab} which is interpreted as the voltage at point **a** with respect to the point **b**. In practice the positive lead of a voltmeter would be placed at **a** and the negative (or common) lead to point **b**.

When a single subscript notation is used to designate the voltage at a point, the voltage reference is assumed to be the chassis common or ground point of the circuit. Given the voltages at two points (V_a and V_b) the voltage difference between the two points can be obtained by the following relationship.

$$V_{ab} = V_a - V_b$$

PROCEDURE

To represent a voltmeter using *Breadboard,* a large resistor ($\geq 1\text{M}\Omega$) is connected between the two points. The large value resistor will not affect the circuit and represents the input resistance of the voltmeter. In Figure 7.1 the resistor of branch 6 is the "voltmeter" and solving the circuit will give the voltage across the two nodes and its polarity.

Draw and solve the circuit shown in Figure 7.1 using the *Breadboard* program. Record the node voltages on the circuit diagram and show the direction of the current in branch 6 and the voltage across branch 6.

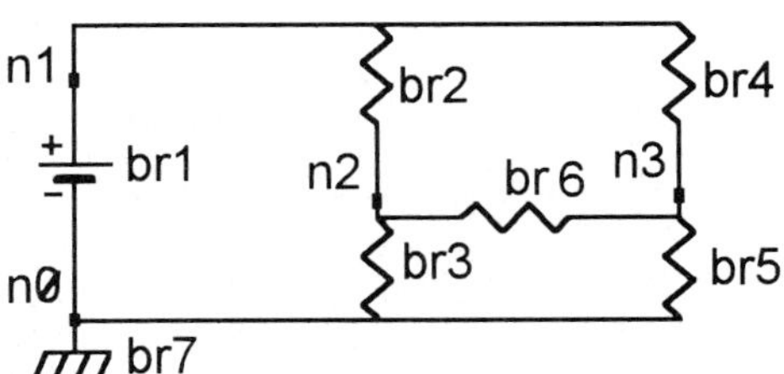

Figure 7.1

Net List for the Circuit

Br. 1: E = 40 V
Br. 2: R = 5 kΩ
Br. 3: R = 5 kΩ
Br. 4: R = 6 kΩ
Br. 5: R = 4 kΩ
Br. 6: R = 10 MΩ
Br. 7: voltage reference

The voltage difference of node 2 with respect to node 3 is

$$V_{ab} = \text{(voltage at node 2)} - \text{(voltage at node 3)} = \underline{\hspace{1cm}} \text{ V}$$

or the voltage across the resistor of branch 6 = ________ V.

PROCEDURE Part 2

Draw and solve the circuit shown in Figure 7.2. Record the node voltages on the circuit diagram.

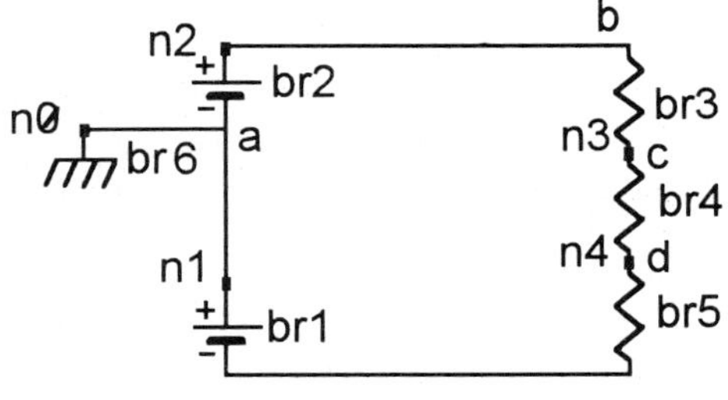

Figure 7.2

Net List for the Circuit

Br. 1: E = 15 V
Br. 2: E = 25 V
Br. 3: R = 2 kΩ
Br. 4: R = 5 kΩ
Br. 5: R = 1 kΩ
Br. 6: voltage reference

Determine the voltage difference (magnitude and polarity) by subtracting the appropriate node voltages.

$$V_{ba} = \underline{\hspace{1.5cm}} \text{ V} \qquad V_{ca} = \underline{\hspace{1.5cm}} \text{ V} \qquad V_{da} = \underline{\hspace{1.5cm}} \text{ V}$$

Confirm your answers by connecting a 10 MΩ resistor between the desired points and solving for the voltage across this resistor.

$$V_{ba} = \underline{\hspace{1.5cm}} \text{ V} \qquad V_{ca} = \underline{\hspace{1.5cm}} \text{ V} \qquad V_{da} = \underline{\hspace{1.5cm}} \text{ V}$$

FURTHER ANALYSIS

Choose a voltage difference problem from your text and use *Breadboard* to confirm your calculations.

Internal Source Resistance

OBJECTIVE

To study the effect of internal source resistance on the terminal resistance of practical voltage sources.

THEORY

Ideal voltage sources are assumed to have no internal resistance. All practical sources, however, have some internal resistance. The effect of internal resistance is to decrease the terminal voltage when current is drawn from the source due to the internal voltage drop. The percent decrease in voltage from no-load to full-load current is called the voltage regulation.

PROCEDURE Part 1

The resistor of branch 2 in the circuit shown in Figure 8.1 represents the internal resistance of the voltage source.

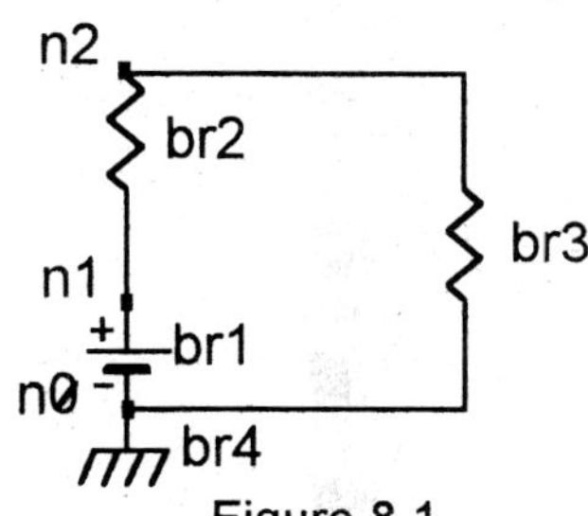

Figure 8.1

Net List for the Circuit

Br. 1: E = 60 V
Br. 2: R = 10 Ω
Br. 3: R = 110 Ω
Br. 4: voltage reference

Draw the circuit using *Breadboard* and solve for the current drawn from the source and the terminal voltage of the source (node 2) for the values of load resistance (branch 3) shown in Table 8.1. Also record the internal voltage drop of the source (branch 2).

Table 8.1

Load Resistance	Source Current branch 1	Terminal Voltage node 2	Internal Voltage Drop
110 Ω			
50 Ω			
20 Ω			
10 Ω			
2 Ω			

PROCEDURE Part 2

The circuit of Figure 8.2 represents a high-voltage DC source with a comparatively high internal resistance. When a low-input resistance voltmeter (branch 4) is used to measure the voltage across the load resistor of branch 3, the current drawn by the voltmeter will load the source and lower its terminal voltage. The reading of the voltmeter will actually be lower than the voltage that existed before adding the meter.

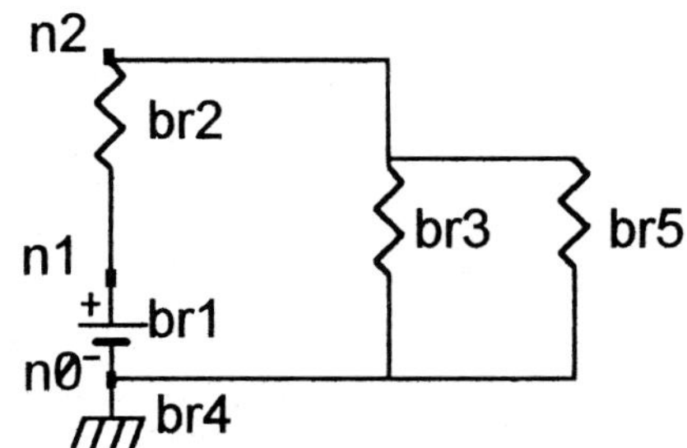

Net List for the Circuit

Br. 1: E = 2400 V
Br. 2: R = 800 Ω
Br. 3: R = 5.6 MΩ
Br. 4: voltage reference
Br. 5: R = 4 kΩ

Figure 8.2

Draw and solve the circuit **without** the voltmeter (branch 5) connected.

Terminal voltage of the source (node 2) = ________ V

Add the branch 5 resistor (voltmeter) to the circuit and solve for the terminal voltage.

Terminal voltage of the source (node 2) with the voltmeter = ________ V

The loading effect of the voltmeter caused the terminal voltage to drop

from ___________ V to ___________ V.

FURTHER ANALYSIS

If a different voltmeter with an input resistance of 10 MΩ is used, the loading effect which drops the terminal voltage of the source would much reduced.

Change the resistor of branch 4 to 10 MΩ and solve the circuit.

Terminal voltage of the source (node 2) with the voltmeter = ________ V

The loading effect of the voltmeter caused the terminal voltage to drop

from ___________ V to ___________ V.

Series-Parallel Circuits

OBJECTIVE

To support the study of the analysis of series-parallel circuits.

THEORY

Most circuits in electronics are combinations of series and parallel connected components. Components are in series if the same current flows through them, and components are in parallel if the voltage across them is the same. To solve and analyze series-parallel circuits the method of network reduction can be used. Parallel components are reduced to their equivalent resistance and series resistances are added until a single total resistance is obtained. The total current is determined and the process is then reversed to find the current and voltage in other parts of the circuit. *Breadboard* will provide the complete solution of series-parallel circuits, eliminating the calculations. To obtain an understanding of circuit operation, however, the results obtained must be analyzed.

PROCEDURE Part 1

For the circuit shown in Figure 9.1, use *Breadboard* to obtain the solution for the current through and voltage across each component.

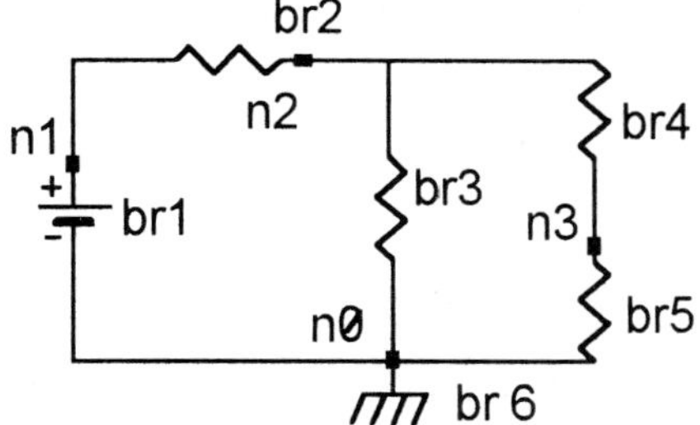

Net List for the Circuit

Br. 1: E = 60 V
Br. 2: R = 1.2 kΩ
Br. 3: R = 3.6 kΩ
Br. 4: R = 1.6 kΩ
Br. 5: R = 2.0 kΩ
Br. 6: voltage reference

Figure 9.1

Record on the circuit diagram the direction and magnitude of the currents, and the polarity and magnitude of the voltage across each resistor.

Prove Kirchhoff's current law at node 2.

At node 2, sum of the currents entering = __________ mA
equals sum of the currents leaving = __________ mA

Prove Kirchhoff's voltage law around each of the meshes.

Mesh 1: sum of voltage rises = _____ V = sum of voltage drops _______ V
Mesh 2: sum of voltage rises = _____ V = sum of voltage drops _______ V

The equivalent resistance of the circuit can be found by dividing the source voltage by the source current.

Equivalent resistance of the circuit = $R_T = E_S/I_S$ = _________ Ω

PROCEDURE Part 2

For the circuit shown in Figure 9.2, solve for the voltages across and current through each of the resistors.

On the circuit diagram, mark the polarity and magnitude of the voltages across and the current through each resistor.

Net List for the Circuit

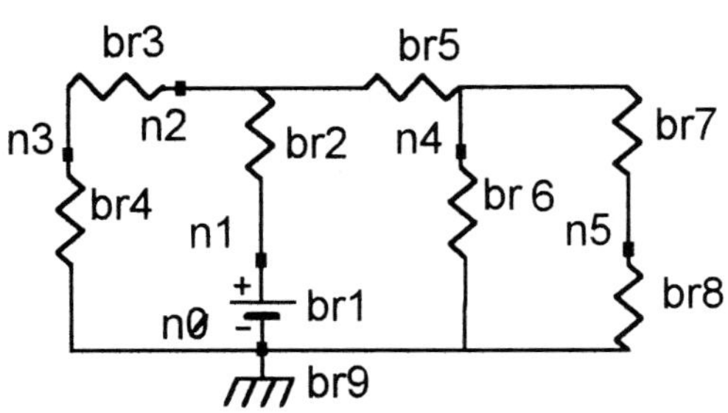

Figure 9.2

Br. 1: E = 42 V
Br. 2: R = 20 Ω
Br. 3: R = 550 Ω
Br. 4: R = 250 Ω
Br. 5: R = 400 Ω
Br. 6: R = 800 Ω
Br. 7: R = 500 Ω
Br. 8: R = 300 Ω
Br. 9: voltage reference

Prove Kirchhoff's current law at node 2.

At node 2, sum of the currents entering = _________ mA
equals the sum of the currents leaving = _________ mA

At node 4, sum of currents entering = _________ mA
equals the sum of currents leaving = _________ mA

For each of the three meshes of the circuit, show that Kirchhoff's voltage law is maintained.

Mesh 1: sum of voltage rises = ______ V = sum of voltage drops ______ V
Mesh 2: sum of voltage rises = ______ V = sum of voltage drops ______ V
Mesh 3: sum of voltage rises = ______ V = sum of voltage drops ______ V

The equivalent resistance of the circuit = $R_T = E_S/I_S$ = _________ Ω

PROCEDURE Part 3

The circuit schematic of Figure 9.3 is called a ladder network due to its configuration. It may also be viewed as a repeated voltage divider circuit. The output voltage from one stage is further reduced by the following voltage divider stage.

Use *Breadboard* to obtain the solution for the voltage at each node and the current in each resistor.

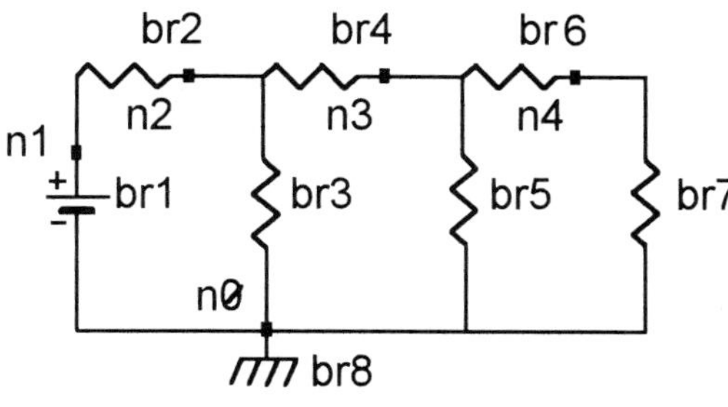

Figure 9.3

Net List for the Circuit

Br. 1: E = 48 V
Br. 2: R = 9 kΩ
Br. 3: R = 27 kΩ
Br. 4: R = 9 kΩ
Br. 5: R = 27 kΩ
Br. 6: R = 9 kΩ
Br. 7: R = 27 kΩ
Br. 8: voltage reference

Record on the circuit diagram the current in each branch and the node voltages.

Prove Kirchhoff's voltage law for the peripheral path of the circuit.

Sum of the voltage rises = _______________ V
equals the sum of the voltage drops _______________ V.

Prove Kirchhoff's current law at node 3.

Sum of the currents entering = ________ mA.
equals the sum of the currents leaving = ________ mA.

PROCEDURE Part 4

Bridge circuits are used in many applications such as measuring instruments and strain gauge configurations. In Figure 9.4 the resistor of branch 7 represents a sensitive galvanometer to indicate zero current (or no voltage difference) when the bridge is balanced. The condition for balance of a resistor bridge is that the product of the resistance of the opposite arms must be equal. An important feature of bridge circuits is that balance is independent of the voltage of the source, eliminating the need for a calibrated voltage source.

If the bridge is unbalanced, the circuit configuration is not a series-parallel circuit and requires Delta-Wye conversion or loop analysis to obtain a solution.

Draw and solve the circuit for an unbalanced condition using the values for the components given in the net list.

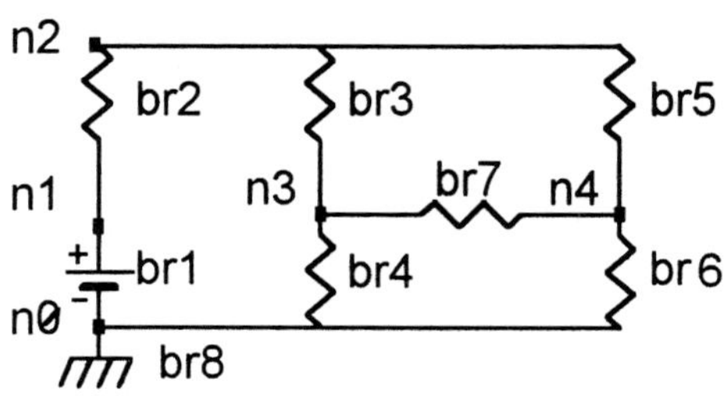

Figure 9.4

Net List for the Circuit

Br. 1: E = 20 V
Br. 2: R = 100 Ω
Br. 3: R = 2.0 kΩ
Br. 4: R = 5.0 kΩ
Br. 5: R = 10.0 kΩ
Br. 6: R = 3.0 kΩ
Br. 7: R = 200 Ω
Br. 8: voltage reference

For the unbalanced bridge the voltage across branch 7 is ___________ V.

The current through the galvanometer (branch 7) is _________ mA.

To balance the bridge change the resistance of branch 6 to 25 kΩ. Solve the circuit and record on the circuit diagram the current in each branch and the voltage across each component.

For the balanced bridge the voltage across branch 7 is ________ V.

The current through the galvanometer is __________ mA.

FURTHER ANALYSIS

Refer to your textbook and select two problems from the section on series-parallel circuits. Solve the problems by the method of network reduction. Then use *Breadboard* to confirm your results.

Voltage Dividers (Loaded)

10

OBJECTIVE

To obtain the effects of load changes on the voltages of a voltage divider circuit.

THEORY

Voltage dividers are used to provide several voltage levels from a single DC source by adding voltage-dropping resistors in series with the source. A typical application would be a multi-stage transistor amplifier. At the rated current of the load the voltage drop in the series resistors decrease the source voltage to the desired level. If the load current drawn from the voltage divider is not constant, an undesirable voltage change at the load terminals may result. To reduce the voltage variation at the load, a bleeder resistor is added to the voltage divider circuit. The larger the bleeder current the more stable the voltage of the divider. A compromise must be made since the bleeder current is wasted power. A bleeder current of at least 10% of the source current may be adequate.

PROCEDURE Part 1

The circuit shown in Figure 10.1 is a simple voltage divider without a bleeder resistor. The current source in branch 3 simulates the current drawn by the load which will be varied to observe the effect on the voltage at node 2. The voltage divider is to provide 20 mA at 10 volts from a 15 volt source. The voltage dropping resistor (branch 2) must then drop 5 volts and have a resistance value of 5 V/20 mA = 250 Ω.

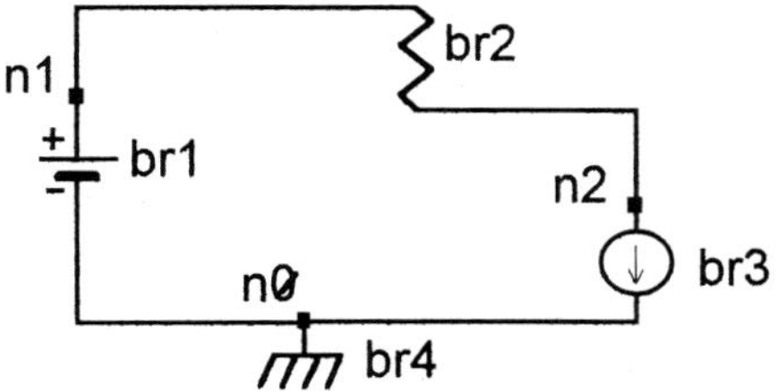

Figure 10.1

Net List for the Circuit

Br. 1: E = 15 V
Br. 2: R = 250 Ω
Br. 3: I = 20 mA
Br. 4: voltage reference

Draw and solve the circuit using *Breadboard.* Complete the corresponding cell in the table. Change the load current (branch 3) to 1 µA and then 40 mA and record the voltage at node 2 in Table 10.1.

Table 10.1

	Without Bleeder	With 10% Bleeder	With 50% Bleeder	With 100% Bleeder
Load Current	Voltage at node 2	Voltage at node 2	Voltage at node 2	Voltage at node 2
1.0 µA				
20 mA				
40 mA				

PROCEDURE Part 2

Modify the circuit diagram by adding a 5 kΩ bleeder resistor (branch 5), as shown in Figure 10.2, which will draw 2 mA or 10% of the load current. Note that the value of the voltage dropping resistor (branch 2) has also been changed to account for the increased current drawn by the bleeder resistor.

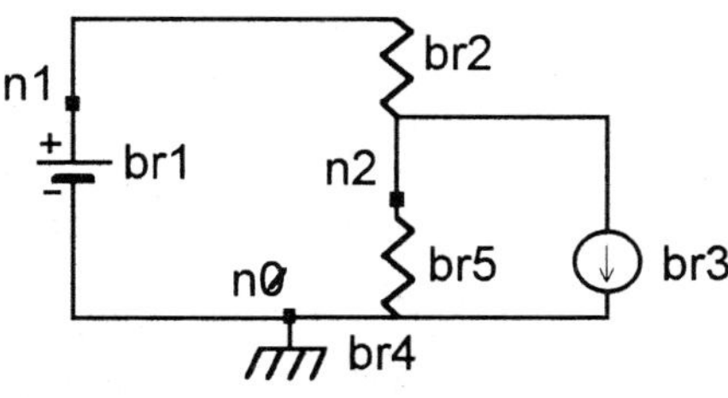

Figure 10.2

Net List for the Circuit

Br. 1: E = 15 V
Br. 2: R = 227 Ω
Br. 3: I = 20 mA
Br. 4: R = 5 kΩ
Br. 5: voltage reference

Solve the circuit and fill in the appropriate cell in Table 10.1.

Change the value of the load current (branch 3) to 1 µA and 40 mA, solve for the voltage at node 2 for each value and fill in the table.

Change the bleeder resistor (branch 5) to 1kΩ to draw 50% of the rated load. Also change the voltage dropping resistor (branch 2) to 167 Ω. Solve for the three load current levels and fill in the table.

Change the bleeder resistor (branch 5) to 500 Ω to draw 100% of the rated load. Also change the voltage dropping resistor (branch 2) to 125 Ω. Solve for the three load current levels and fill in the table.

Study the tabulated results and comment on the voltage variation with load current changes as obtained with different bleeder current values.

Comments

Power in DC Circuits

OBJECTIVE

To confirm the calculation of power dissipated in a resistor and the total power in a circuit.

THEORY

Power in a resistor is given by the formula:

$$P = I^2R = V^2/R = VI$$

where I is the current through the resistor and V is the voltage across the resistor.

The total power in the circuit is the sum of the power in each of the resistors. Since the power in the circuit must be provided by the source(s), the total power can also be obtained by the sum of the power delivered by the sources. The power delivered by a source is the source voltage times the current leaving the source. If the current be entering the positive terminal of a source, the source is absorbing power as, for example, a battery under charge.

PROCEDURE

Draw and solve the circuit in Figure 11.1 for the component values shown in the net list.

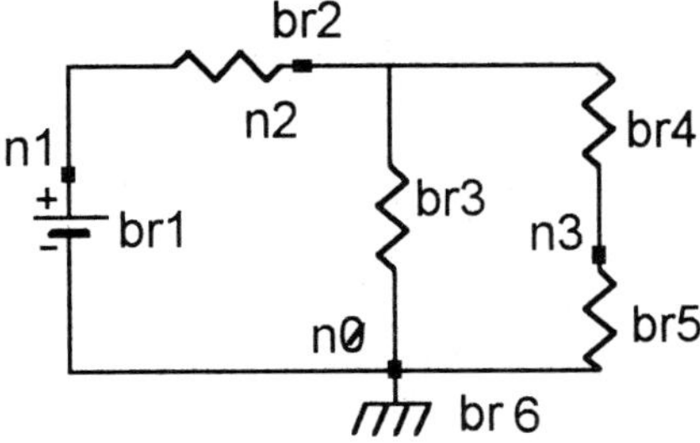

Figure 11.1

Net List for the Circuit

Br. 1: E = 125 V
Br. 2: R = 50 Ω
Br. 3: R = 400 Ω
Br. 4: R = 100 Ω
Br. 5: R = 300 Ω
Br. 6: voltage reference

From the *Breadboard* solution results of branch voltage and branch current, calculate the power in each branch and enter the values in the table below. Enter the *Breadboard* solution for the power in the table to confirm your calculations.

Table 11.1

Branch	R	I	V	I^2R	IV	V^2/R	Power
2	50 Ω						
3	400 Ω						
4	100 Ω						
5	300 Ω						

Total power dissipated in the resistors = _____________________ W

Calculate the following:

Total current from the source = I_S = _________ A

Total power from the source (EI_S) = _________ W

Total equivalent resistance of the circuit $(R_T = E/I_S)$ = _________ Ω

Total power delivered (E^2/R_T) = _________ W

Total power delivered $(I_S^2R_T)$ = _________ W

Breadboard solution for total power is _________ W

FURTHER ANALYSIS

Repeat the above procedure for the circuit shown in Figure 11.2. The circuit has two power sources connected series aiding. Each source is then providing power to the circuit.

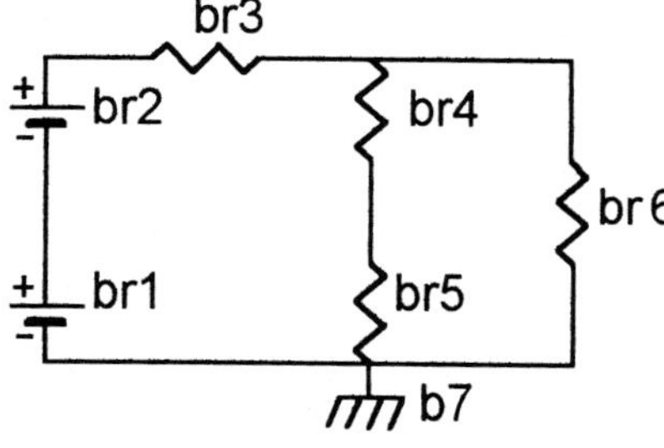

Figure 11.2

Net List for the Circuit

Br. 1: E = 20 V
Br. 2: E = 10 V
Br. 3: R = 1 kΩ
Br. 4 R = 2.7 kΩ
Br. 5: R = 1.8 kΩ
Br. 6: R = 5.6 kΩ
Br. 7: voltage reference

Reverse the polarity of the voltage source in branch 2. Reconcile the power delivered and absorbed by the components in the circuit.

Ideal Constant-Current Sources

OBJECTIVE

To become familiar with constant-current sources and to solve circuits that include constant-current sources.

THEORY

Constant-current sources are the dual of constant-voltage sources. Although constant-current sources do not naturally occur, they are available as power supplies where the supply provides constant current by current-regulating electronic circuitry. The concept of a constant-current source is used in the nodal analysis method of circuit analysis.

A constant-current source delivers the specified value of current to the network connected to it regardless of the resistance of the network. The voltage across the constant-current source will depend on the resistance of the network. By comparison, the voltage of a constant-voltage source is fixed and the current supplied by a constant-voltage source is dependent on the resistance of the network connected to it.

Just as an ideal constant-voltage source does not have internal resistance, an ideal constant-current source does not have an internal resistor connected in parallel.

PROCEDURE

Using *Breadboard*, solve the circuit shown in Figure 12.1.

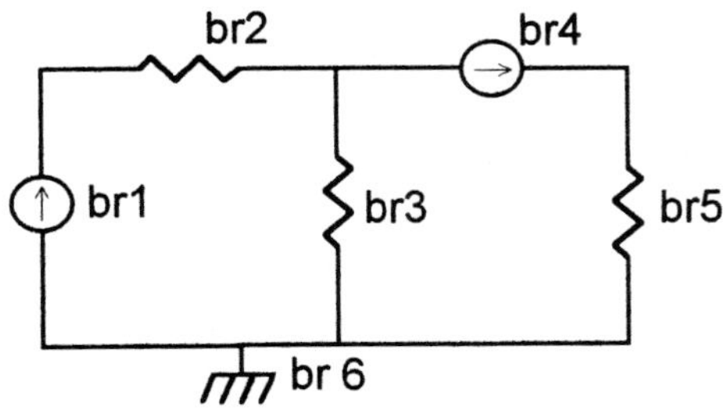

Net List for the Circuit

Br. 1: I = 2 A
Br. 2: R = 4 Ω
Br. 3: R = 10 Ω
Br. 4: I = 3 A
Br. 5: R = 6 Ω
Br. 6: voltage reference

Figure 12.1

Mark on the circuit diagram the branch voltages and the current (magnitude and direction) for each branch.

Confirm Kirchhoff's current law at node 2 of the circuit.

Node 2: current entering _________A = current leaving ______A

What is the voltage across the 2 A current source? $V_{branch\ 1}$ = _________ V

What is the voltage across the 3 A current source? $V_{branch\ 4}$ = _________ V

Confirm Kirchhoff's voltage law for each mesh

Mesh 1: Sum of the voltage rises = ______ V
                    equals sum of the voltage drops ______V

Mesh 2: Sum of the voltage rises = ______ V
                    equals sum of the voltage drops ______V

FURTHER ANALYSIS

Using *Breadboard*, solve the circuits in Figures 12.2 and 12.3 and analyze the circuits as in the procedure above.

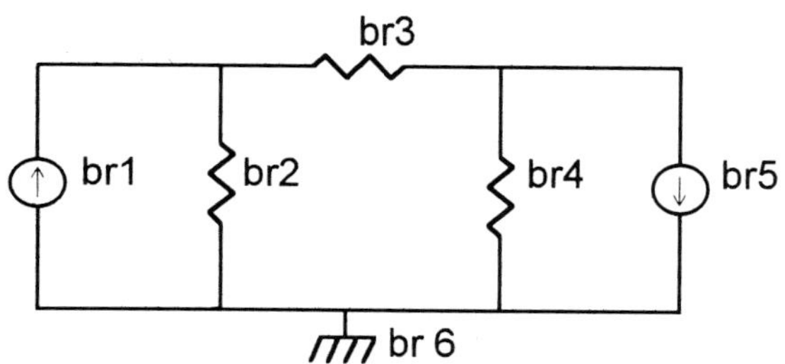

Figure 12.2

Net List for the Circuit

Br. 1: I = 5 A
Br. 2: R = 4 Ω
Br. 3: R = 12 Ω
Br. 4: R = 3 Ω
Br. 5: I = 6 A
Br. 6: voltage reference

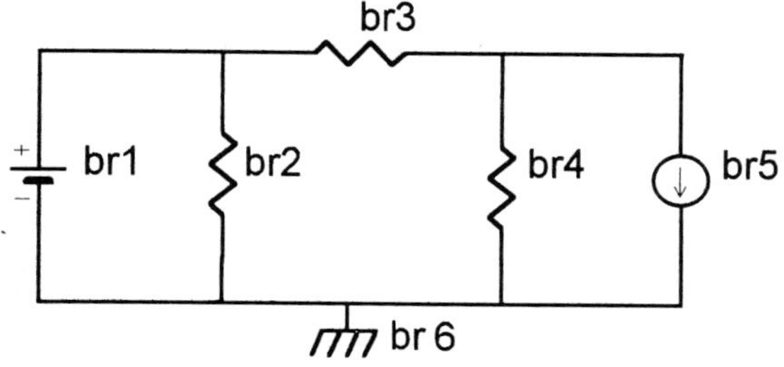

Figure 12.3

Net List for the Circuit

Br. 1: E = 20 V
Br. 2: R = 4 Ω
Br. 3: R = 6 Ω
Br. 4: R = 8 Ω
Br. 5: I = 4 A
Br. 6: voltage reference

Constant-Current, Constant-Voltage Source Duality

OBJECTIVE

To confirm the equivalency of a constant-current source and a constant-voltage source.

THEORY

If a resistor (R_S), representing the internal resistance of the constant-current source of magnitude I_S, is connected in parallel with the source, the **terminal voltage characteristics** of the current source will be identical to a constant-voltage source whose voltage magnitude is I_S times R_S and whose internal series resistance is R_S. The terminal voltage (V_{ab}) provided by the current source will equal the terminal voltage of the voltage source for all values of load resistance.

The internal power loss of the constant-current source is fictional, since at no load, all the current of the source flows in the internal resistance. This is unlike the voltage source, where an internal power loss occurs only when the voltage source is delivering current to the load.

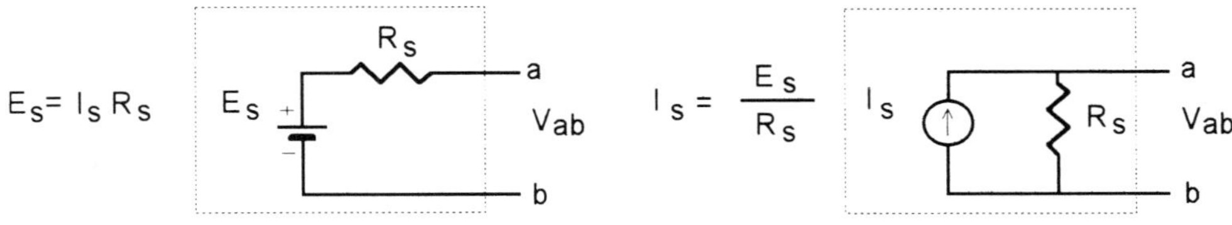

Figure 13.1

PROCEDURE

The load resistor (branch 3) of the circuits in Figures 13.2 and 13.3 represents the total resistance of a network connected to the two types of sources. Using *Breadboard*, draw and solve the circuits shown to prove that the current through the load resistor (and therefore the voltage across the resistor) is the same for each type of source.

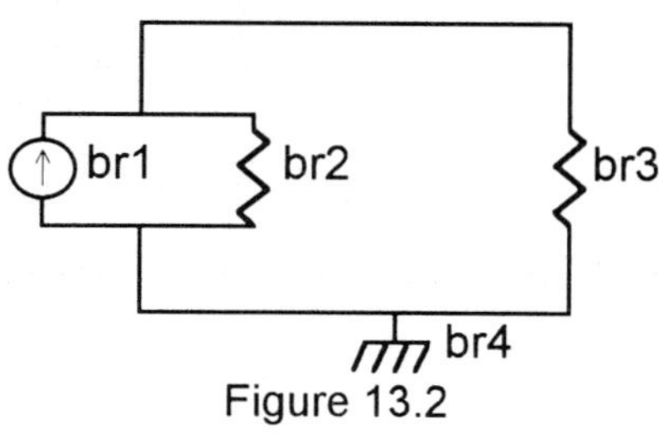

Figure 13.2

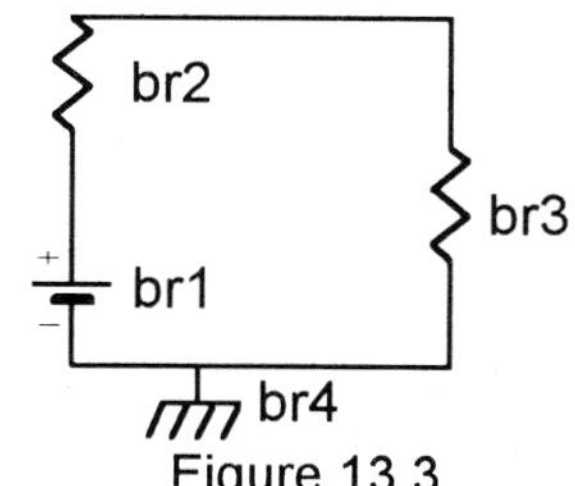

Figure 13.3

Net List for the Circuit

Br. 1: I = 500 mA
Br. 2: R = 20 Ω
Br. 3: R = 80 Ω
Br. 4: voltage reference

Net List for the Circuit

Br. 1: E = 10 V
Br. 2: R = 20 Ω
Br. 3: R = 80 Ω
Br. 4: voltage reference

Solve the two circuits using *Breadboard* and complete Table 13.1.

Table 13.1

Load Resistance	Current Source		Voltage Source	
	Load Current	Load Voltage	Load Current	Load Voltage
2 Ω				
20 Ω				
200 Ω				
2 kΩ				
20 kΩ				

Is the terminal voltage the same for the current source and the voltage source?

What would be the open-circuit terminal voltage of
         the current source? _______________ V
         the voltage source? _______________ V

What would be the short-circuit current of
         the current source? _______________ A
         the voltage source? _______________ A

FURTHER ANALYSIS

Using *Breadboard's* graphing feature, obtain graphs of the current through the load resistor for the two types of sources shown in Figures 13.2 and 13.3 for the range of load resistance 1 Ω to 100 kΩ. From the graphs, compare the current through the load for several values of load resistance. Are they equal?

Circuits with More than One Source

OBJECTIVE

To analyze DC circuits containing more than one source.

THEORY

When voltage sources with their inherent internal resistance are connected in parallel, the method of network-reduction cannot be applied to obtain a solution. The analysis requires the methods of mesh or nodal analysis or the application of Thevenin's or Norton's theorem. Computer analysis programs such as *Breadboard* greatly facilitate obtaining a solution from which an analysis can be made.

PROCEDURE Part 1

The circuit in Figure 14.1 represents an automobile battery (branch 1) and generator (branch 4) with their respective internal resistances connected to a load. For the conditions shown, is the generator supplying the load and charging the battery, or are both the generator and the battery supplying load current?

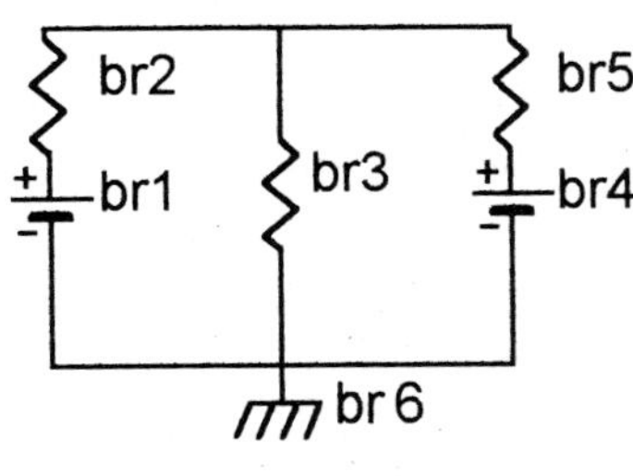

Figure 14.1

Net List for the Circuit

Br. 1: E = 12.6 V
Br. 2: R= 0.4 Ω
Br. 3: R = 1.0 Ω
Br. 4: E = 14.2 V
Br. 5: R = 0.4 Ω
Br. 6: voltage reference

Draw and solve the circuit of Figure 14.1 using *Breadboard*.

Record the magnitude and direction of the branch currents on the circuit diagram. If the current is entering the positive terminal of the battery then the battery is being charged and power is absorbed by that source.

The battery (branch 1) is _________________ (charging or discharging).

The terminal voltage of the two sources (node 2) is ____________ V.

The power supplied by the two sources must be equal to the power dissipated in all the resistors.

The total power absorbed is ___________ W.

The total power delivered by the sources is ___________ W.

PROCEDURE Part 2

Repeat the procedure above for the voltage conditions of the battery (branch 1) shown in Figure 14.2.

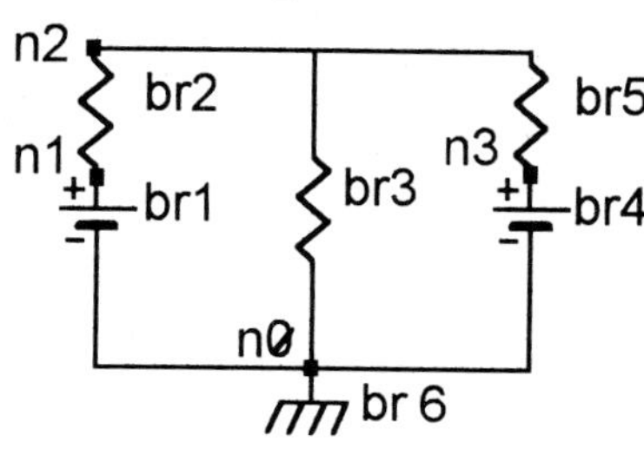

Figure 14.2

Net List for the Circuit

Br. 1: E = 9.8 V
Br. 2: R = 0.5 Ω
Br. 3: R = 1.0 Ω
Br. 4: E = 14.2 V
Br. 5: R = 0.4 Ω
Br. 6: voltage reference

Is the battery (branch 1) charging or discharging? _______________

The terminal voltage of the two sources (node 2) is _______________ V.

The power supplied must be equal to the power absorbed and dissipated.

The total power absorbed is _______________ W.

The total power delivered by the source(s) is ___________ W.

PROCEDURE Part 3

Power sources are connected in parallel whenever one source is incapable of supplying the required load current. Two DC generators are represented in Figure 14.3 as a generated voltage (branches 1 and 4) and their internal resistances (branches 2 and 5). The terminal voltage (node 2) of the two generators must be the same by virtue of the parallel connection. If the generated voltages and internal resistances are identical or proportional to one another, load sharing will also be equal or proportional to their power ratings.

If the generators are not identical in electrical characteristics the current magnitude and direction of each generator will be such that the voltage drop across their internal resistances will cause the terminal voltage to be the same. If the generated voltages are considerably different, one generator may actually absorb power and run as a motor. The condition in each generator is not readily apparent unless an analysis of the circuit is performed.

Draw and solve the circuit of Figure 14.3.

Mark on the circuit diagram the magnitude and direction of the current in each branch and the voltage at node 2.

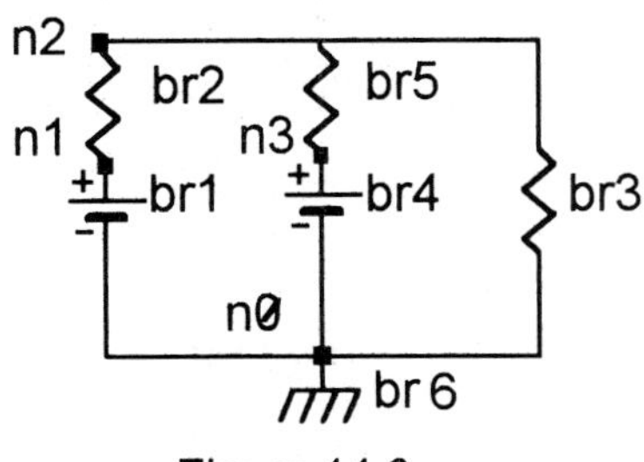

Figure 14.3

Net List for the Circuit

Br. 1: E = 120 V
Br. 2: R = 0.5 Ω
Br. 3: R = 4 Ω
Br. 4: E = 115 V
Br. 5: R = 0.4 Ω
Br. 6: voltage reference

Power delivered is the power from the sources (branches 1 and 4) _________ W.

Power dissipated in the resistance branches (2, 3, and 5) is _______________ W.

PROCEDURE Part 4

Change the generated voltage of the generator in branch 4 to 100 volts and solve the circuit.

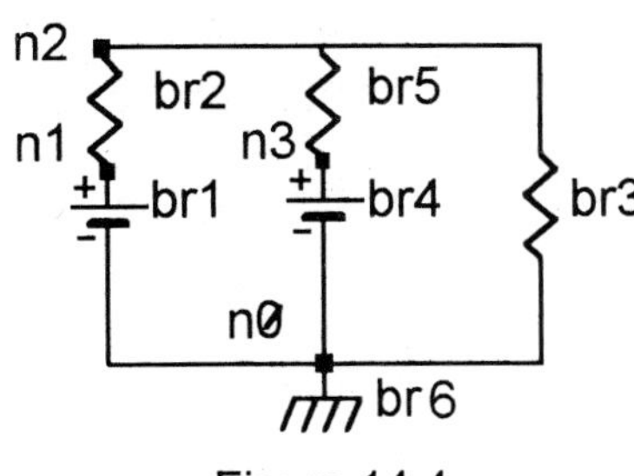

Figure 14.4

Net List for the Circuit

Br. 1: E = 120 V
Br. 2: R = 0.5 Ω
Br. 3: R = 4 Ω
Br. 4: E = 100 V
Br. 5: R = 0.4 Ω
Br. 6: voltage reference

Mark on the circuit diagram the magnitude and direction of the current in each branch and the voltage at node 2.

Is the generator depicted by branch 4 and 5 delivering power or is it "motoring"?

____

Power delivered is the power from the source in branch 1 _________ W.

Power absorbed by the source in branch 4 is _______________ W.

Power dissipated in the resistance branches (2, 3, and 5) is _________ W.

Total power delivered _______ W equals total power absorbed _______W.

FURTHER ANALYSIS

A two-voltage distribution system is depicted in Figure 14.5. The resistances of branches 3, 5, and 7 represent the resistances of the distribution lines. Solve the circuit using *Breadboard*. Record on the circuit diagram the current in each branch.

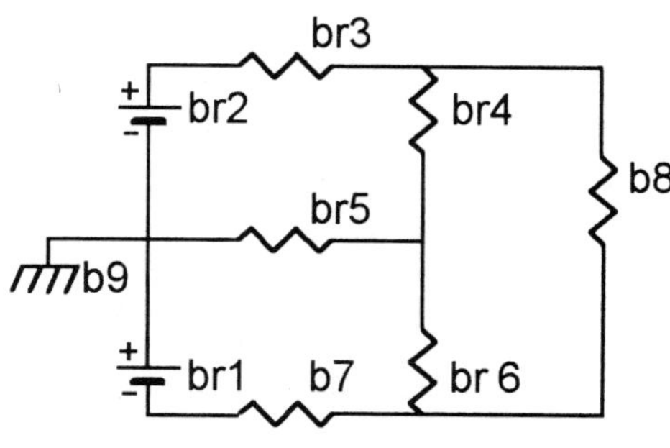

Figure 14.5

Net List for the Circuit

Br. 1: E = 110 V
Br. 2: E = 110 V
Br. 3: R = 0.1 Ω
Br. 4: R = 10 Ω
Br. 5: R = 0.2 Ω
Br. 6: R = 12 Ω
Br. 7: R = 0.1 Ω
Br. 8: R = 22 Ω
Br. 9: voltage reference

Confirm Kirchhoff's current law at nodes 3, 4, and 5.

at node 3, current entering = _______ A equals currents leaving _______ A

at node 4, current entering = _______ A equals currents leaving _______ A

at node 5, current entering = _______ A equals currents leaving _______ A

Comment on the value of current in the neutral wire (branch 5) of the three-wire system.

Comment

Thevenin's Theorem

DC Exercise

PURPOSE

To use *Breadboard* to facilitate obtaining the Thevenin equivalent of a complex circuit.

THEORY

Although *Breadboard* obviates the need for Thevenizing a circuit, the concept of Thevenin's theorem is important to understanding circuit operation. The theorem states that any linear two terminal network may be replaced by an equivalent circuit consisting of a constant voltage source in series with a resistance. In effect the equivalent circuit is just a practical voltage source.

The procedure for obtaining the Thevenin voltage and resistance requires the following steps.

1. Remove the load resistor and determine the voltage that exists across the open terminals. This voltage is the Thevenin voltage.

2. Determine the Thevenin resistance by replacing all voltage sources with a short circuit and all current sources with an open circuit, and obtain the resistance of the circuit as would be measured between the open terminals.

PROCEDURE Part 1

The Thevenin equivalent for the circuit shown in Figure 15.1 is to be obtained and the current through the resistor of branch 5 is to be determined. Solve the circuit using *Breadboard*, and record on the circuit diagram the magnitude and direction of the current in branch 5.

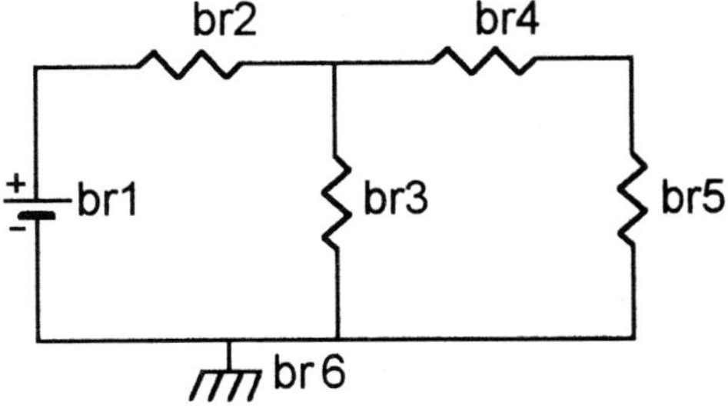

Net List for the Circuit

Br. 1: E = 30 V
Br. 2: R = 4 kΩ
Br. 3: R = 4 kΩ
Br. 4: R = 2 kΩ
Br. 5: R = 2 kΩ
Br. 6: voltage reference

Figure 15.1

To obtain the Thevenin voltage, remove the load resistor (branch 5) and replace it with a 10 MΩ resistance (representing a voltmeter). Solve the circuit and obtain the voltage across the resistor. This is the Thevenin voltage.

The Thevenin voltage = E_{TH} = ____________ V

To obtain the Thevenin resistance, R_{TH}, replace the voltage source (branch 1) with a connecting wire. Replace the 10 MΩ resistor (branch 5) with a 10 V voltage source (to act as an ohmmeter battery), and solve the circuit. The Thevenin resistance is the 10 V divided by the current delivered by this source.

The Thevenin resistance = R_{TH} = 10 V/I_S ____________ Ω

The equivalent Thevenin circuit is shown in Figure 15.2. It consists of the Thevenin voltage source in series with the Thevenin resistance and the original load resistance (1.8 kΩ) connected to the terminals. Draw and solve the Thevenin equivalent circuit of Figure 15.2. Determine the current in the load resistor of branch 3.

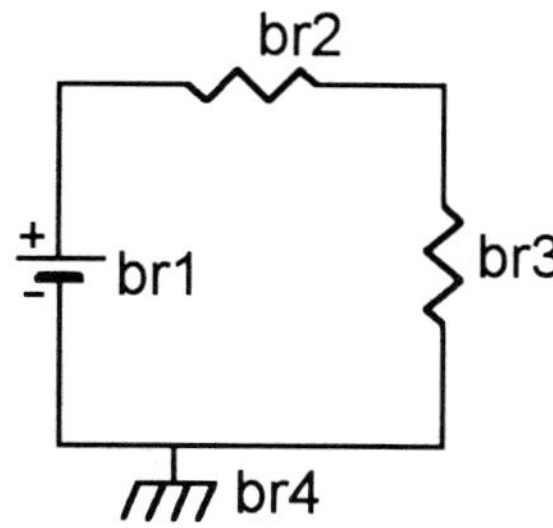

Figure 15.2

Net List for the Circuit

Br. 1: E = Thevenin voltage E_{TH}
Br. 2: R = Thevenin resistance R_{TH}
Br. 3: R = 2 kΩ (load resistance)
Br. 4: voltage reference

Current in the load resistor (branch 3) = ____________ mA

Is the current in the load resistor the same as that of the original circuit? ______

PROCEDURE Part 2

In the circuit of Figure 15.3, the current in the load resistor of branch 4 is to be determined using the method of Thevenin's theorem. Solve the circuit using *Breadboard* and record on the circuit diagram the magnitude and direction of the current in branch 4.

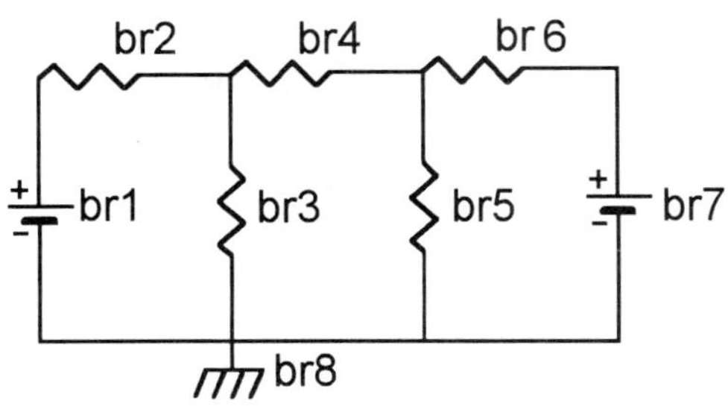

Figure 15.3

Net List for the Circuit

Br. 1: E = 20 V
Br. 2: R = 800 Ω
Br. 3: R = 1.2 kΩ
Br. 4: R = 300 Ω
Br. 5: R = 1.8 kΩ
Br. 6: R = 1.2 kΩ
Br. 7: E = 15 V
Br. 8: voltage reference

Following the procedure above, determine the Thevenin voltage and resistance.

The Thevenin voltage (E_{TH}) = ___________ V

The Thevenin resistance (R_{TH}) = 10 V/I_S ___________ Ω

Draw and solve the equivalent Thevenin circuit shown in Figure 15.4 and determine the current through the load resistor.

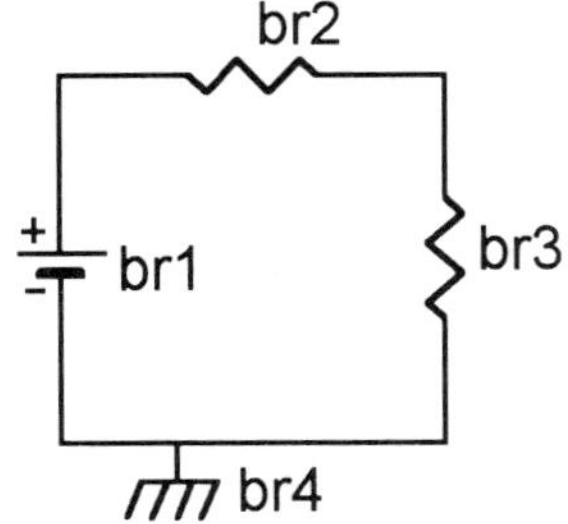

Figure 15.4

Net List for the Circuit

Br. 1: E = Thevenin voltage E_{TH}
Br. 2: R = Thevenin resistance R_{TH}
Br. 3: R = 1.8 kΩ (load resistance)
Br. 4: voltage reference

The current through the load resistor (branch 3) is ___________ mA.

Do the results confirm the equivalence of the Thevenin circuit? ___________

PROCEDURE Part 3

An alternative method of determining the Thevenin voltage and resistance is to remove the load resistor and determine the open circuit voltage and the short circuit current that would flow when the open terminals are shorted.

1. Find the open circuit voltage (E_{oc}) at the load resistor by removing the load resistor and replacing it with a very high value resistor (representing a voltmeter): for example, 10 MΩ.

2. Find the short-circuit current (I_{sc}) by replacing the load resistor with a very low value resistor (representing an ammeter): for example, 0.0001Ω.

The Thevenin voltage is equal to E_{oc} and the Thevenin resistance equals E_{oc}/I_{sc}.

Using *Breadboard* solve the unbalanced bridge circuit of Figure 15.5 and record on the circuit diagram the current through branch 7.

Net List for the Circuit

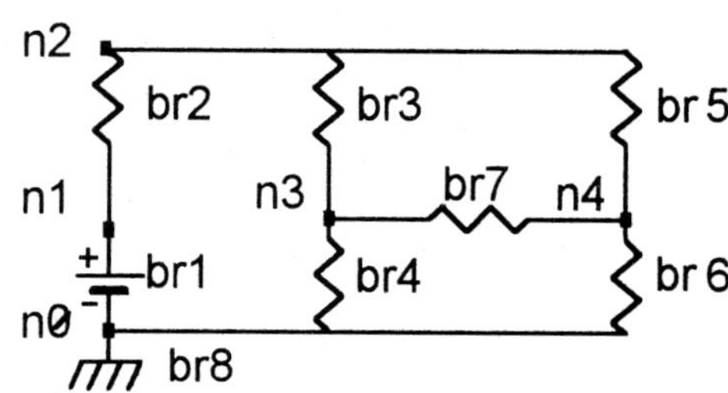

Figure 15.5

Br. 1: E = 20 V
Br. 2: R = 100 Ω
Br. 3: R = 2.0 kΩ
Br. 4: R = 5.0 kΩ
Br. 5: R = 1.0 kΩ
Br. 6: R = 4.0 kΩ
Br. 7: R = 200 Ω
Br. 8: voltage reference

Replace the resistor of branch 7 with a 10 MΩ resistor and solve the circuit. The voltage across this resistor is the open circuit voltage E_{oc}.

The open circuit voltage (E_{oc}) = ______________ V.

Replace the resistor of branch 7 with a 0.0001 Ω resistor representing a short circuit and solve for the current (I_{sc}).

The short circuit current (I_{sc}) = ____________ mA.

The Thevenin resistance (R_{TH}) = E_{oc}/I_{sc} = ____________ Ω.

Using *Breadboard* draw the Thevenin equivalent circuit (as in Figure 15.4) and solve for the current in the load resistor.

Does the load current obtained for the Thevenin equivalent circuit equal the original load current? _________

Norton's Theorem

16

PURPOSE

To use *Breadboard* to facilitate obtaining the Norton equivalent circuit of a complex network.

THEORY

Norton's theorem is the dual of Thevenin's theorem. Norton's theorem states that any two-terminal linear circuit can be replaced by an equivalent circuit consisting of a current source paralleled by a resistor. In other words, the equivalent circuit is just a practical current source.

To obtain the value of the equivalent current source and resistance, the following procedure is used.

     1. Disconnect the load resistor in which the current is to be determined and find the short circuit current (I_N) that would flow between the terminals.

     2. Replace all voltage sources with a short circuit and all current sources with an open circuit. Determine the resistance (R_N) looking back into the open terminals.

The Norton equivalent circuit is the current source (I_N) paralleled with its internal resistance (R_N) and the original load resistance connected to the terminals.

PROCEDURE

Draw and solve the circuit to be Nortonized shown in Figure 16.1. Mark on the circuit diagram the current in the load resistor (branch 5).

Net List for the Circuit

Br. 1: E = 20 V
Br. 2: R = 0.4 Ω
Br. 3: R = 0.8 Ω
Br. 4: E = 28 V
Br. 5: R = 1.0 Ω
Br. 6: voltage reference

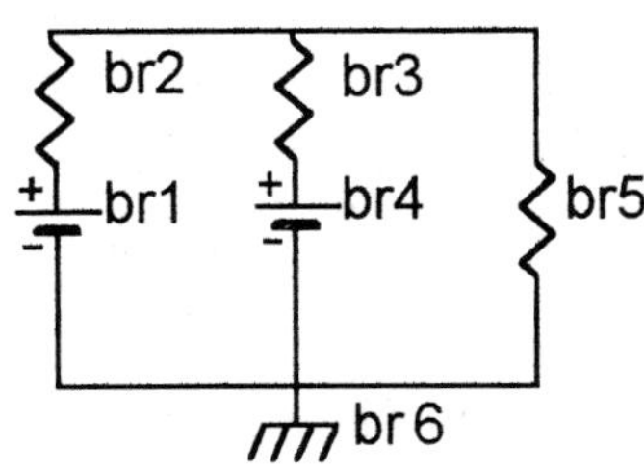

Figure 16.1

To find the Norton current (I_N), replace the load resistor (branch 5) with a very small resistance (e.g. 0.001 Ω) to represent an ammeter acting as a short circuit between the open terminals. Solve the circuit and obtain the current through the shorting resistor.

The short circuit current = I_N =_________ A.

Edit the circuit and replace the source voltages (branch 1 and branch 4) with a connecting wire. Replace the resistor in branch 5 with a 10 volt source. The Norton resistance is the 10 volts divided by the current delivered by this source. Solve for the current from this source

The Norton resistance is 10 V/I_S = ____________ Ω.

Draw the Norton equivalent circuit shown in Figure 16.2 and solve for the current through the original load resistor (branch 3).

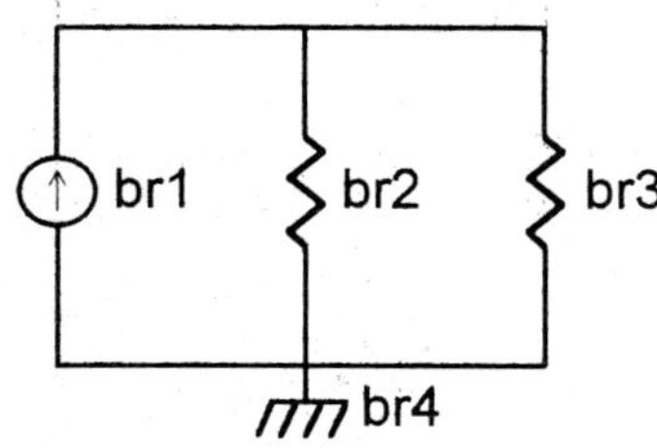

Figure 16.2

Net List for the Circuit

Br. 1: I = Norton current source I_N
Br. 2: R = Norton resistance R_N
Br. 3: R = load resistance = 1.0 Ω
Br. 4: voltage reference

The current through branch 3 is __________ A.

Does the load current in the Norton equivalent circuit equal the original load current? ________

FURTHER ANALYSIS

Follow the above procedure and determine the equivalent Norton circuit for the network shown below. Using *Breadboard* solve the original circuit and the Norton equivalent for the current in the resistor of branch 4.

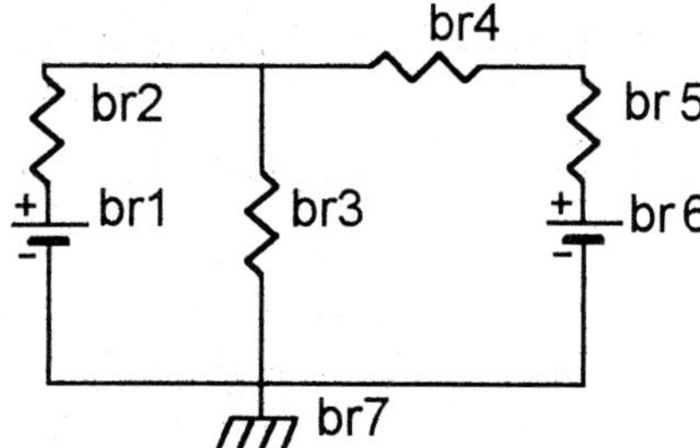

Figure 16.3

Net List for the Circuit

Br. 1: E = 60 V
Br. 2: R= 10 Ω
Br. 3: R = 50 Ω
Br. 4: R = 40 Ω
Br. 5: R = 2 Ω
Br. 6: E = 40 V
Br. 7: voltage reference

Maximum Power Transfer

PURPOSE

To confirm the theorem of maximum power transfer.

THEORY

In accordance with the theorem, maximum power is delivered to the load when the load resistance equals the internal resistance of the source. When the two resistances are equal, the power loss in the internal resistance is equal to the power in the load resistor. This efficiency of 50% is acceptable for electronic circuits where maximum power transfer is more important than efficiency and the quantities of power are very small. In power systems, however, efficiency takes preference and the load resistance is not matched to the internal resistance of the source.

PROCEDURE

Since any complex circuit of passive elements can be represented as a single resistor by circuit-reduction methods or Thevenin's theorem, the resistor in branch 2 of the circuit in Figure 17.1 may be considered to be the total resistance of any complex circuit or just the internal resistance of the source. The resistor in branch 3 represents the load resistance of the original circuit in which the power is to be maximized.

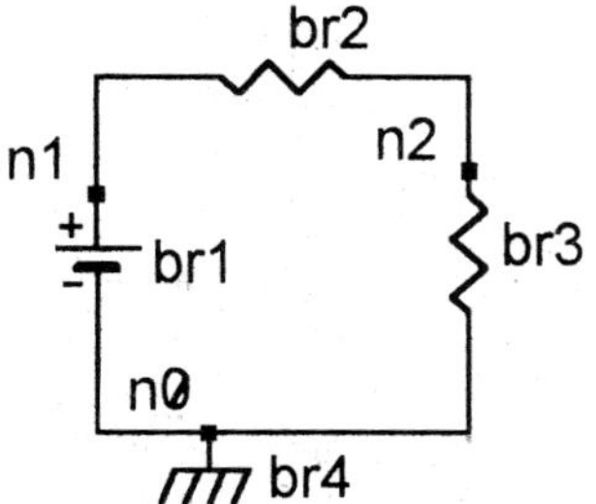

Figure 17.1

Net List for the Circuit

Br. 1: E = 40 V
Br. 2: R = 100 Ω
Br. 3: R = 100 Ω
Br. 4: voltage reference

Draw the circuit for the values given in the net list using *Breadboard*.

Select the graph option from the menu and obtain the graph of power in the load resistor (branch 3) as the value of the load resistance varies from 1 Ω to 1000 Ω.

From the graph, obtain the value of maximum power and the value of load resistance at which it occurs.

The maximum power transferred is _______ W for a load resistance of _______ Ω.

At maximum power does the resistance of branch 3 equal that of branch 2? _____

Obtain a graph of the voltage output (node 2) for values of load resistance (branch 3) from 1 Ω to 1000 Ω.

The value of output voltage at which maximum power is obtained is _______ V.

FURTHER ANALYSIS

According to the theorem, maximum power will be delivered to the load resistor when its value equals the Thevenin's resistance of the circuit. The circuit in Figure 17.2 is a potential divider circuit with the resistor of branch 5 as the load resistor.

Determine the Thevenin resistance of the circuit. $R_{TH} = $ _________ Ω

Obtain a graph of the power in the load resistor as the value of the resistor varies from 100 Ω to 10 kΩ.

Compare the value of load resistance for maximum power transfer, obtained from the graph, to the calculated Thevenin resistance.

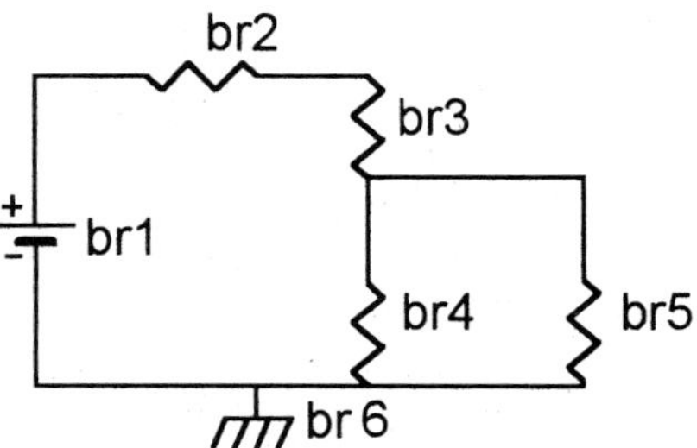

Figure 17.2

Net List for the Circuit

Br. 1: E = 80 V
Br. 2: R = 500 Ω
Br. 3: R = 500 Ω
Br. 4: R = 1000 Ω
Br. 5: R = 1000 Ω
Br. 6: voltage reference

For the Norton equivalent circuit shown in Figure 17.3, obtain a graph of the power in the load resistor (branch 3) as the value of the load resistor varies from 100 Ω to 10 kΩ. For maximum power transfer, is the resistance of branch 3 equal to that of branch 2?

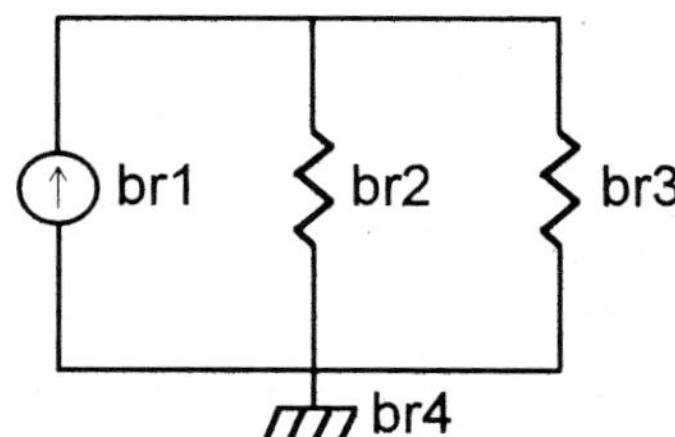

Figure 17.3

Net List for the Circuit

Br. 1: I = 20 mA
Br. 2: R = 1000 Ω
Br. 3: R = 1000 Ω
Br. 4: voltage reference

Current-Controlled Current Source

OBJECTIVE

To investigate the characteristics of a current-controlled current source.

THEORY

A current-controlled current source (CCCS) delivers a current whose magnitude is determined by the current flowing in some other branch of the circuit. The magnitude of the current from the CCCS is dependent on the current flowing in the controlling branch multiplied by the amplification factor (BETA) of the CCCS. This type of controlled source is used to model a BJT transistor where the base current of the transistor multiplied by the BETA ($\beta = I_C/I_B$) determines the current that flows in the collector.

For the circuit of Figure 18.1, the current in the resistor of the controlling branch (branch 2) causes a current BETA times as large to flow from the CCCS in branch 4 regardless of the resistance of branch 3 in series with the CCCS.

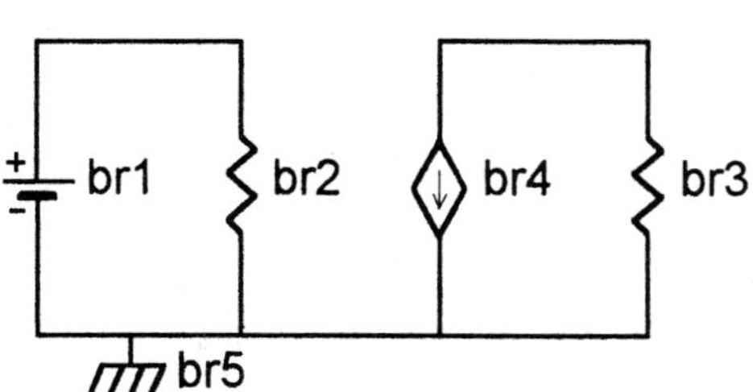

Figure 18.1

Net List for the Circuit

Br. 1: E = 10 mV
Br. 2: R = 100 Ω
Br. 3: R = 1 kΩ
Br. 4: CCCS, beta = 80
        control branch 2
Br. 5: voltage reference

PROCEDURE Part 1

Using *Breadboard*, draw the circuit and obtain the solution for the circuit. Record the results in Table 18.1. Mark the values of current and voltage on the circuit diagram of Figure 18.1.

Note: The direction of the movement of the cursor just prior to placing the control branch resistor must be downward (for Figure 18.1) to provide the program with the current direction in the control resistor. Similarly, for the CCCS, the cursor movement must be downward prior to drawing the CCCS. Generally, it is good practice to draw the components in a sequence that follows the direction of conventional current through them.

Using the editing feature of *Breadboard* (deleting and replacing a component), change the value of the voltage source in branch 1 to the remaining values in Table 18.1. Solve and complete the table.

Table 18.1

E branch 1	Control current branch 2	CCCS current branch 4	BETA $I_{branch\ 4}/I_{branch\ 2}$	Voltage branch 4
100 µV				
1.0 mV				
10 mV				
100 mV				

The calculated BETA for each value of input voltage is ___________ .

Note: The voltage across the CCCS (branch 4) represents the collector-to-emitter voltage of a transistor. This voltage is important in establishing the operating point of a transistor. See exercises on transistor biasing.

PROCEDURE Part 2

For an input voltage of 10 mV in the circuit of Figure 18.1, change the BETA of the CCCS to 80, 100, 150, and 200 and solve the circuit. Determine that the current gain of the CCCS is BETA times the current in branch 2. Complete table 18.2.

Table 18.2

BETA	Control current branch 2	CCCS current branch 4	Calculated $I_{branch\ 4}$ $= \beta\ (I_{branch\ 2})$	Voltage branch 4
80				
100				
150				
200				

FURTHER ANALYSIS

Obtain a graph of the current in branch 3 as the resistance of branch 3 is varied from 100 Ω to 10 kΩ. Comment on the graphical results with reference to the characteristics of a CCCS.

Voltage-Controlled Current Source 19

PURPOSE

To study the action of a voltage-controlled current source in a simple DC circuit.

THEORY

The current magnitude supplied by a voltage-controlled current source (VCCS) is dependent on the voltage across some other resistor in the circuit multiplied by a gain factor called the mutual transconductance (G_m) of the VCCS. The G_m of the device is defined as the current output of the VCCS divided by the voltage across the controlling resistor and has the units of Siemens (formerly mhos, the reciprocal of ohms). This type of dependent source is used to model a FET transistor where the gate-source voltage determines the current that flows from drain to source of the FET.

PROCEDURE

In Figure 19.1 the magnitude of the current provided by the VCCS in branch 4 is determined by the voltage drop across the resistor of branch 3 multiplied by the transconductance (G_m) of the VCCS, specified as 2000 µS.

Net List for the Circuit

Br. 1: E = 10 mV
Br. 2: R = 1 MΩ
Br. 3: R = 1 MΩ
Br. 4: VCCS, G_m = 2000 µS
    control branch 3
Br. 5: R = 5 kΩ
Br. 6: voltage reference

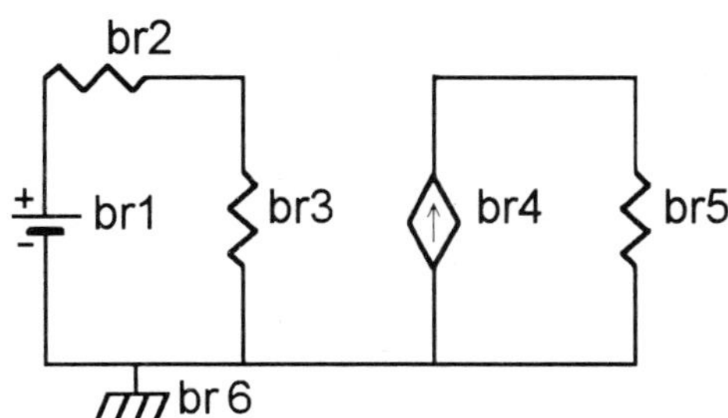

Figure 19.1

Using *Breadboard*, draw the circuit of Figure 19.1 and solve. From the solution results, mark on Figure 19.1 the current and voltage of each branch. Enter the required values in Table 19.1.

Note: Draw the components in the sequence that follows the conventional current flow through them. The branch number sequence in the net list indicates the order in which to draw the components.

Repeat the above procedure for each of the remaining input voltages listed in the table.

Table 19.1

E branch 1	Control voltage branch 3	VCCS current branch 4	G_m $I_{branch\,4}/V_{branch\,3}$	Voltage branch 5
1 mV				
10 mV				
100 mV				
1.0 V				

For the 10 mV input voltage, is the current in branch 4 equal to the voltage of branch 3 multiplied by the transconductance of the VCCS?

$$I_{branch\,4} = V_{branch.\,3} \cdot G_m = \underline{\hspace{2cm}} \ A$$

$$\textit{Breadboard} \text{ solution for } I_{branch\,4} = \underline{\hspace{2cm}} \ A$$

With reference to the results obtained, comment on and explain the characteristics of a VCCS.

Comment

FURTHER ANALYSIS

For an input voltage of 10 mV, draw and solve the circuit of Figure 19.1 for values of G_m of 2000 µS, 2500 µS, 4000 µS, and 5000 µS. Prove by calculation that the current in branch 5 equals the transconductance times the voltage across branch 3.

Table 19.2

G_m µS	Control voltage branch 3	VCCS current branch 4	G_m $I_{branch\,4}/V_{branch\,3}$	$I_{branch\,5} =$ $G_m \cdot V_{branch\,3}$
2000				
2500				
4000				
5000				

Voltage-Controlled Voltage Source **20**

OBJECTIVE

To study the characteristics of a voltage-controlled voltage source in a simple DC circuit.

THEORY

The voltage developed by a voltage-controlled voltage source (VCVS) is dependent on the voltage across the controlling resistor in the circuit. The magnitude of the voltage of the VCVS is the value of the controlling voltage multiplied by the voltage amplification factor (A_V) of the VCVS. The VCVS is used to model circuits containing voltage amplification, particularly the integrated circuit operational amplifier.

PROCEDURE

In the circuit of Figure 20.1, the voltage across the control branch (branch 3) is amplified by the VCVS in branch 4. Using *Breadboard*, draw and solve the circuit for the values in the net list. On the circuit diagram, mark the voltages and currents of each component. Enter the values in the table.

Note: The sequence of drawing the components must follow the direction of current in them.

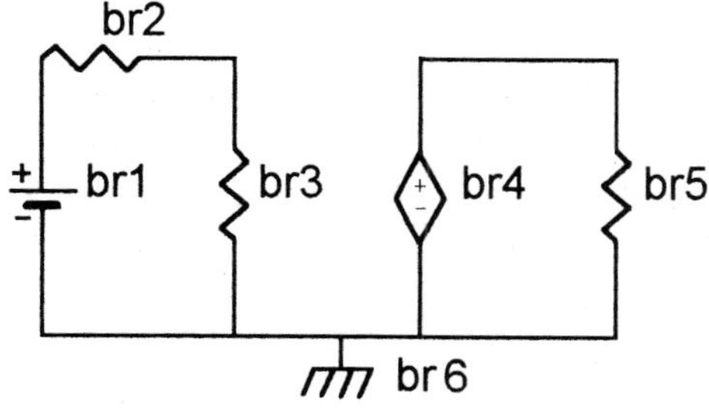

Figure 20.1

Net List for the Circuit

Br. 1: E = 20 mV
Br. 2: R = 1 MΩ
Br. 3: R = 1 MΩ
Br. 4: VCVS, A_V = 10000
      control branch 3
Br. 5: R = 1 kΩ
Br. 6: voltage reference

Using the editing feature of *Breadboard,* modify the circuit for the remaining values of voltage input given in Table 20.1, solve, and complete the table. The calculated value of A_V is $V_{branch. 4}/V_{branch. 3}$.

Table 20.1

E branch 1	Control voltage branch 3	VCVS volts branch 4	A_V $V_{branch 4}/V_{branch 3}$	Current $I_{branch 5}$
1 mV				
2 mV				
4 mV				
8 mV				

Is the voltage gain A_V in Table 20.1 equal to that specified for the VCVS?

$$A_V = V_{branch. 5}/V_{branch. 3} = \underline{\hspace{3cm}}$$

In your own words explain the operation of the VCVS.

FURTHER ANALYSIS

Modify the circuit of Figure 20.1 for values of voltage amplification A_V of the VCVS of 10,000, 20,000, 50,000 and 100,000. Solve the circuit and complete Table 20.2. Does the gain in voltage equal the A_V of the amplifier ?

Table 20.2

A_V	Control voltage branch 3	VCVS volts branch 4	A_V $V_{branch 4}/V_{branch 3}$	Current $I_{branch 5}$
10,000				
20,000				
50,000				
100,000				

Inductive Reactance

OBJECTIVE

To study the characteristics of inductors in AC circuits.

THEORY

The opposition to current flow of an inductor (inductive reactance) is given by the equation $X_L = 2\pi f L$ ohms. The magnitude of the inductive reactance is directly proportional to the frequency (f) of the source in Hertz and to the value of the inductance (L) in Henrys. The current that flows through the inductor lags the voltage across the inductor by 90°.

A practical inductor always has the resistance of the wire making up the coil associated with its inductance. If the magnitude of the inductive reactance is ten times the magnitude of the resistance or more, the resistance can usually be ignored in circuit analysis. Otherwise the inductor can be modeled as a resistor in series with the inductor.

PROCEDURE Part 1

The circuit shown in Figure 1.1 shows an ideal inductor connected to an AC source.

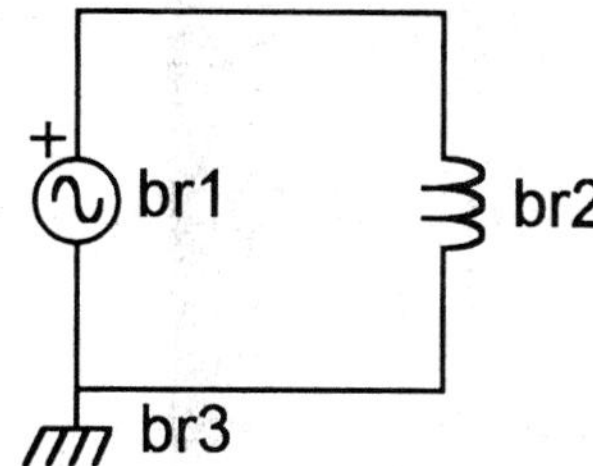

Net List for the Circuit

Br. 1: E = 5 V/0°
Br. 2: L = 40 mH
Br. 3: voltage reference

Figure 1.1

To study the effect of frequency on the magnitude of the inductive reactance, solve the circuit for the frequencies shown in Table 1.1 and complete the table.

Table 1.1

Frequency	I	Phase Angle	$X_L = V_L/I_L$
10 Hz			
100 Hz			
1 kHz			
10 kHz			
100 kHz			

Does the variation of the inductive reactance X_L increase as the frequency is increased? __________

PROCEDURE Part 2

The total opposition to current flow for inductors connected in series is additive.

$$X_T = X_1 + X_2 + X_3 + \dots + X_N$$

The equivalent inductance of inductors connected in series is the sum of the individual inductances.

$$L_T = L_1 + L_2 + L_3 + \dots + L_N$$

For the circuit shown in Figure 1.2, solve the circuit using *Breadboard* for a frequency of 10 kHz and mark on the circuit diagram the current and the voltages across each inductor.

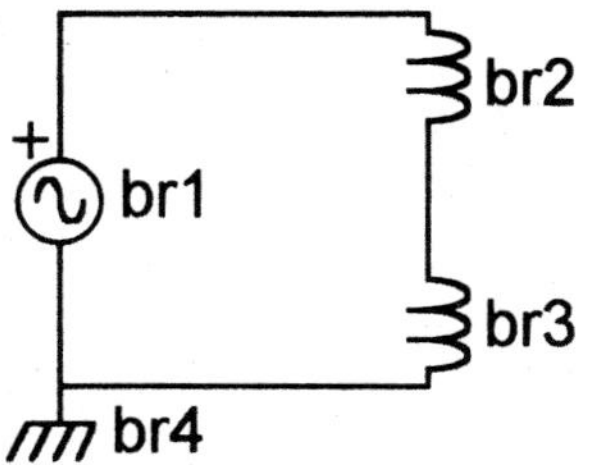

Net list for the Circuit

Br. 1: E =10/0°
Br. 2: L =10 mH
Br. 3: L = 20 mH
Br. 4: voltage reference

Figure 1.2

The inductive reactance of the 10 mH coil = $X_1 = V_L/I_L =$ __________ Ω.

The inductive reactance of the 20 mH coil = $X_2 = V_L/I_L =$ __________ Ω.

The total inductive reactance of the circuit = $X_1 + X_2 =$ __________ Ω.

Calculated total inductive reactance of the circuit is $E/I =$ __________ Ω.

The equivalent inductance of the circuit is $L_1 + L_2 =$ __________ mH.

From the total inductive reactance calculate the equivalent total inductance as

$$L_{eq} = 1/(2\pi f X_T) =$$ __________ mH.

PROCEDURE Part 3

The total inductive reactance of inductors in parallel is obtained by the same relationship as resistors in parallel.

$$\frac{1}{X_{eq}} = \frac{1}{X_1} + \frac{1}{X_2} + \frac{1}{X_3} + \cdots + \frac{1}{X_N}$$

Since $X_L = 2\pi f L$ the equivalent total inductance of inductors in parallel is

$$\frac{1}{L_{eq}} = \frac{1}{L_1} + \frac{1}{L_2} + \frac{1}{L_3} + \cdots + \frac{1}{L_N}$$

Using *Breadboard*, solve for the current through the inductors in the circuit shown in Figure 1.2 at a frequency of 10 kHz.

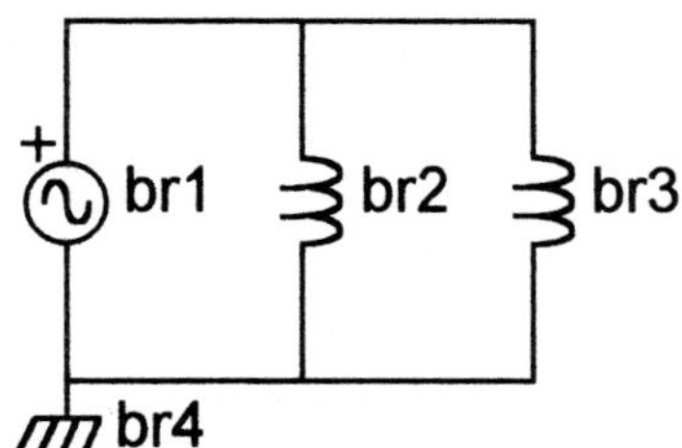

Net List for the Circuit

Br. 1: E = 10 V/0°
Br. 2: L = 20 mH
Br. 3: L = 40 mH
Br. 4: voltage reference

Figure 1.3

The inductive reactance of the 20 mH coil = $X_1 = V_L/I_L =$ _______________ Ω.

The inductive reactance of the 40 mH coil = $X_2 = V_L/I_L =$ _______________ Ω.

Calculated total inductive reactance $X_T = X_1 X_2/(X_1 + X_2) =$ _______________ Ω.

The total inductive reactance of the circuit is $E/I_T =$ _______________ Ω.

The equivalent inductance of the circuit is $L_1 L_2/(L_1 + L_2) =$ _______________ mH.

From the total inductive reactance, calculate the equivalent total inductance as

$$L_{eq} = X_T/(2\pi f) = \text{\_\_\_\_\_\_\_\_\_\_\_\_\_\_\_\_} \text{ mH}.$$

PROCEDURE Part 4

A practical inductor has some resistance associated with its inductance due to the resistance of the wire making up the coil. Depending on the size and length of the wire and the frequency at which the coil is to be used, the resistance may be significant.

A figure of merit for the inductor, called the Q_L of the coil, is expressed as the ratio of the inductive reactance to its resistance.

$$Q_L = X_L/R_L = 2\pi f L/R_L$$

The higher the Q_L the better the inductor and Q_L values of 10 to 200 are common. Although the definition of Q_L indicates that Q_L will increase directly as the frequency increases other effects tend to maintain the Q_L constant over a range of frequencies. These effects are the increase in AC resistance due to the skin effect and the reduction of the effective inductance due to the distributed capacitance of the coils.

The circuit in Figure 1.3 represents a practical inductor with its coil resistance connected to an AC source. Solve the circuit using *Breadboard* for the frequencies shown in Table 1.2 and complete the table.

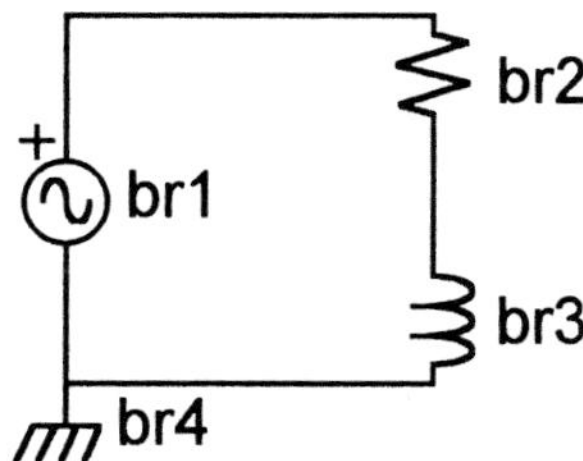

Figure 1.4

Net List for the Circuit

Br. 1: E = 10 V/0°
Br. 2: R = 50 Ω
Br. 3: L = 20 mH
Br. 4: voltage reference

Table 1.2

frequency	I	V_L	$X_L = V_L/I$	$Q_L = X_L/R_1$	$Z = E/I$
100 Hz					
1 kHz					
10 kHz					

Comment on the value of Q and the phase angle of the current as the frequency is increased.

Comment

FURTHER ANALYSIS

Use *Breadboard* to obtain a graph of the current in the circuit of Figure 1, over the frequency range of 1.0 kHz. to 1.0 MHz. Write your comments below on the variation of current as it relates to the reactance of the component.

Comment

Capacitive Reactance

OBJECTIVE

To study the characteristics of capacitors in AC circuits.

THEORY

The opposition to current flow of a capacitor (capacitive reactance) is given as $X_C = 1/(2\pi f C)$ ohms. The magnitude of the capacitive reactance is inversely proportional to the frequency (f) and to the magnitude of the capacitor (C) in Farads. Thus the higher the frequency, the smaller the opposition to current flow. The current that flows leads the voltage across the capacitor by 90°.

PROCEDURE Part 1

The circuit shown in Figure 2.1 shows an ideal capacitor connected to an AC source.

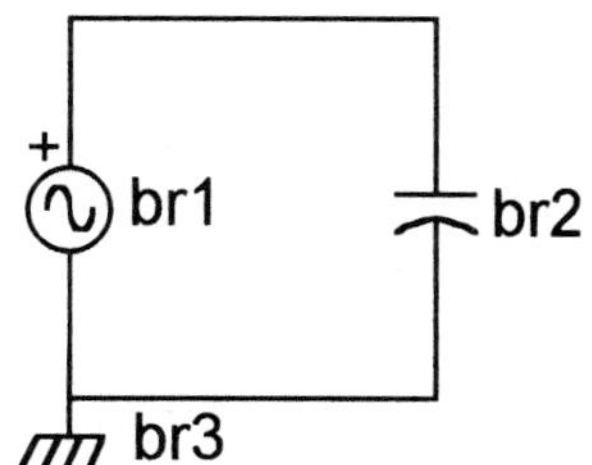

Figure 2.1

Net List for the Circuit

Br. 1: E = 1.0 V/0°
Br. 2: C = 0.01 μF
Br. 3: voltage reference

To study the effect of frequency on the magnitude of the capacitive reactance, solve the circuit using *Breadboard* for the frequencies shown in Table 2.1 and complete the table.

Table 2.1

Frequency	I	Phase Angle	$X_C = V_C/I_C$
10 Hz			
100 Hz			
1 kHz			
10 kHz			
100 kHz			

Does the magnitude of the capacitive reactance vary inversely as the frequency?

PROCEDURE Part 2

The total opposition to current flow for capacitors connected in series is the sum of the individual reactances.

$$X_T = X_1 + X_2 + X_3 + \ldots + X_N$$

The equivalent capacitance of several capacitors connected in series is given by the equation below.

$$\frac{1}{C_{eq}} = \frac{1}{C_1} + \frac{1}{C_2} + \frac{1}{C_3} + \cdots + \frac{1}{C_N}$$

For the circuit shown in Figure 2.2, solve the circuit using *Breadboard* for a frequency of 10 kHz and mark on the circuit diagram the current and the voltage across each capacitor.

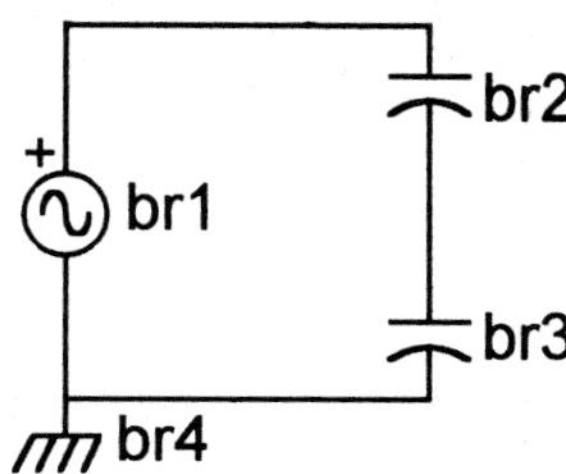

Net List for the Circuit

Br. 1: E = 20 V/0°
Br. 2: C = 0.02 µF
Br. 3: C = 0.04 µF
Br. 4: voltage reference

Figure 2.2

The capacitive reactance of the 0.02 µF capacitor = $X_1 = V_C/I_C$ = __________ Ω.

The capacitive reactance of the 0.04 µF capacitor = $X_2 = V_C/I_C$ = __________ Ω.

Calculated total capacitive reactance $X_T = X_1 + X_2$ = __________ Ω.

The total capacitive reactance of the circuit is E/I = __________ Ω.

The equivalent capacitance of the circuit is $C_1 C_2 /(C_1 + C_2)$ = __________ µF.

From the total capacitive reactance calculate the equivalent total capacitance as

$$C_{eq} = 1/(2\pi f X_T) = \underline{\hspace{3cm}} \ \mu F.$$

PROCEDURE Part 3

The total reactance of capacitors connected in parallel is obtained by the same relationship as resistors in parallel.

$$\frac{1}{X\mathbf{eq}} = \frac{1}{X_1} + \frac{1}{X_2} + \frac{1}{X_3} + \cdots + \frac{1}{X_N}$$

The equivalent capacitance of capacitors connected in parallel is the sum of the individual capacitors.

$$C\mathrm{eq} = C_1 + C_2 + C_3 + \cdots + C_N$$

Using *Breadboard* solve for the current through each capacitor shown in the circuit of Figure 2.3 at a frequency of 10 kHz.

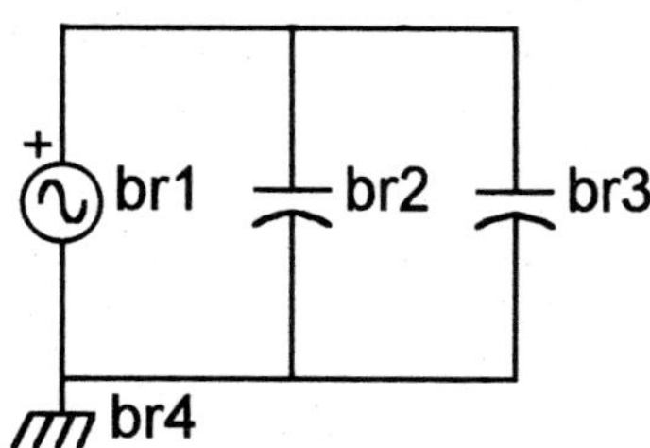

Net List for the Circuit

Br. 1: E = 20 V/0°
Br. 2: C = 0.02 µF
Br. 3: C = 0.04 µF
Br. 4: voltage reference

Figure 2.3

The capacitive reactance of the 0.02 µF capacitor = $X_1 = V_C/I_C =$ _______ Ω

The capacitive reactance of the 0.04 µF capacitor = $X_2 = V_C/I_C =$ _______ Ω.

Calculated total capacitive reactance $X_T = X_1 X_2/(X_1 + X_2) =$ _________ Ω.

The total capacitive reactance of the circuit is $E/I_T =$ _____________ Ω.

The equivalent capacitance of the circuit is $C_1 + C_2 =$ ___________ µF.

From the total capacitive reactance calculate the equivalent total capacitance as

$$C_{\mathrm{eq}} = 1/(2\pi f X_T) = \underline{\hspace{4cm}} \ \mu F.$$

PROCEDURE Part 4

A practical capacitor having some dielectric leakage current can be modeled as a capacitance (C_p) paralleled by a high value resistance (R_p). A figure of merit for a capacitor called the dissipation factor (D) is defined as the reactance of the capacitor divided by the resistance representing the leakage path.

$$D = X_C/R_P$$

The smaller the value of the dissipation factor the better is the capacitor, and D factors of 0.0001 are typical.

The circuit of Figure 2.4 represents a practical capacitor. Solve the circuit for a frequency of 50 kHz and mark the current of each branch on the circuit diagram.

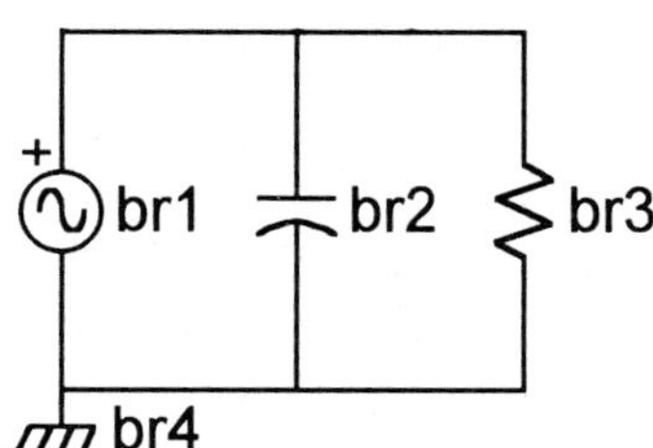

Net List for the Circuit

Br. 1: E = 20 V/0°
Br. 2: C = 0.1 µF
Br. 3: R = 1 MΩ
Br. 4: voltage reference

Figure 2.4

The reactance of the capacitor $X_C = E/I_C = $ ________________ Ω.

The dissipation factor $D = X_C/R = $ ________________ .

FURTHER ANALYSIS

Use *Breadboard* to obtain a graph of the current in the capacitor of Figure 2.1 over the frequency range of 1.0 kHz to 100 kHz. Write your comments below on the variation of current as it relates to the reactance of the capacitor.

Comment

Series *R-L* and *R-C* Circuits 3

OBJECTIVE

To determine the voltages and currents in R-L and R-C circuits, and to obtain the impedance and phasor diagrams.

THEORY

In an AC series R-L or R-C circuit the current is common to all components, but the voltages across the inductance or the capacitance are 90° out of phase with the current and the voltage across the resistor is in phase with the current. In order to calculate the current magnitude and phase angle, the method of complex algebra is used. An impedance diagram facilitates the calculation of the impedance of the circuit.

A phasor diagram is used to relate the current and voltages of the circuit in a graphical form.

PROCEDURE Part 1

For an R-L series circuit as shown in Figure 3.3 the current is calculated as E/Z_T where both E and Z are complex numbers having both magnitude and phase angle. In an R-L circuit the current will lag the applied voltage.

The impedance $Z = R + jX_L = \sqrt{R^2 + X_L^2} \angle \tan^{-1}\left(\dfrac{X_L}{R}\right)$

The impedance diagram of Figure 3.1 shows the relationship of R, X_L, and Z and the phasor diagram of Figure 3.2 shows the relationship of the current and the voltages across the components.

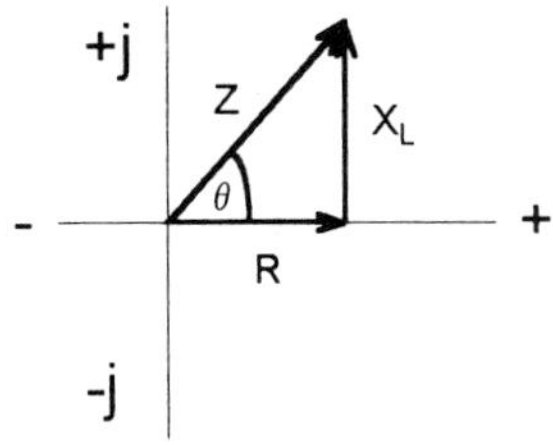

Figure 3.1

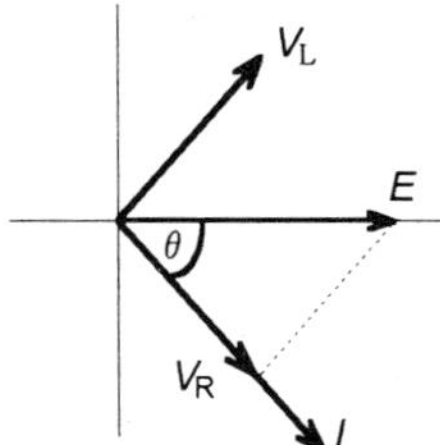

Figure 3.2

Using *Breadboard* draw and solve the *R-L* circuit shown in Figure 3.3 for a frequency of 60 Hz.

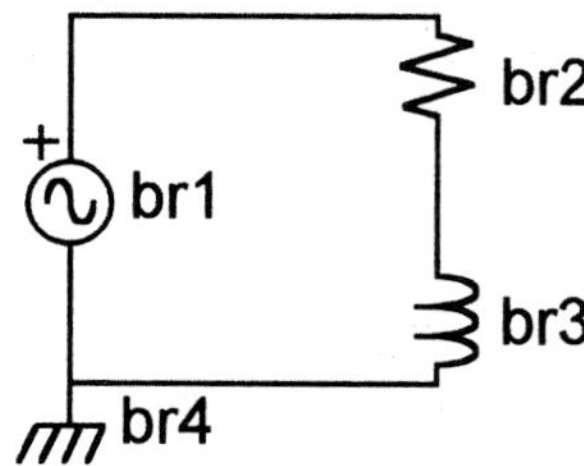

Figure 3.3

Net List for the Circuit

Br. 1: $E = 120$ V/0°
Br. 2: $R = 100\ \Omega$
Br. 3: $L = 0.5$ H
Br. 4: voltage reference

On the circuit diagram mark the current and the voltages across each component.

From the solutions obtained using *Breadboard*, calculate X_L and Z_T.

$$X_L = V_L/I = \underline{\hspace{3cm}} \angle \underline{\hspace{2cm}}°\ \Omega$$

$$Z_T = E/I = \underline{\hspace{3cm}} \angle \underline{\hspace{2cm}}°\ \Omega$$

Draw the impedance diagram and the phasor diagram for the circuit approximately to scale.

Confirm Kirchhoff's voltage law on the phasor diagram by graphically summing the voltage across the resistor and the inductor.

$$E = V_R + V_L$$

PROCEDURE Part 2

To determine the impedance of an *R-C* circuit, complex algebra must be used to sum the resistance and capacitive reactance to obtain the impedance of the circuit.

$$Z = R - jX_C = \sqrt{R^2 + X_C^2} \angle \tan^{-1}\left(\frac{-X_C}{R}\right)$$

The current in the circuit will lead the applied voltage by the angle of the impedance. Figure 3.4 and 3.5 are the impedance and phasor diagrams for an *R-C* circuit.

Figure 3.4   Figure 3.5

Draw and solve the circuit shown in Figure 3.6 for a frequency of 1.0 MHz.

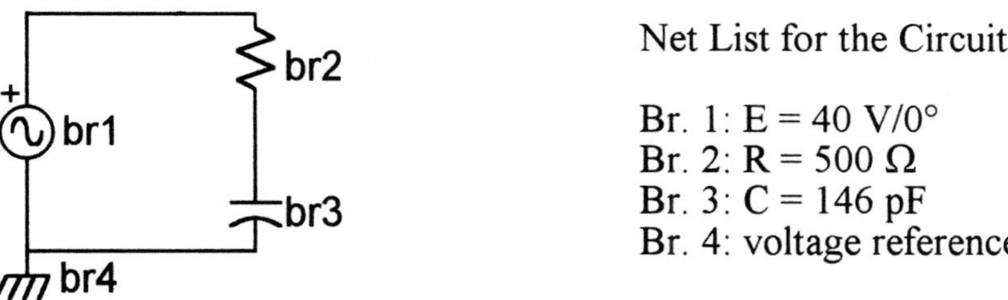

Figure 3.6

Net List for the Circuit

Br. 1: E = 40 V/0°
Br. 2: R = 500 Ω
Br. 3: C = 146 pF
Br. 4: voltage reference

On the circuit diagram mark the current and the voltages across the resistor and the capacitor.

From the solution results, calculate the capacitive reactance and the impedance of the circuit.

$$X_C = V_C/I = \underline{\hspace{3cm}} \angle \underline{\hspace{3cm}}° \; \Omega$$

$$Z_T = E/I = \underline{\hspace{3cm}} \angle \underline{\hspace{3cm}}° \; \Omega$$

What is the phase angle of the current? ________ °

Draw the impedance diagram and the phasor diagram for E, I, V_C, and V_R.

Impedance Diagram   Phasor Diagram

Confirm Kirchhoff's voltage law on the phasor diagram by graphically summing the voltage across the resistor and the inductor.

$$E = V_R + V_C$$

FURTHER ANALYSIS

Refer to your textbook and choose a problem from the section on *R-L* and *R-C* series circuits. Repeat the procedure above for drawing the phasor and impedance diagrams.

Series *R-L-C* Circuits

OBJECTIVE

To obtain the solution of the current and the voltages in a series *R-L-C* circuit, and to calculate the impedance of the circuit at two different frequencies.

THEORY

In a series *R-L-C* circuit, the current is common to all components. The voltage across the resistor is in phase with the current, whereas the voltage across the inductor and the capacitor respectively lead and lag the current by 90°. In applying Kirchhoff's voltage law, the voltages must be added as phasors.

The impedance of an *R-L-C* circuit requires the algebra of complex numbers and is given by the equation below.

$$Z = R + j(X_L - X_C) = \sqrt{R^2 + (X_L - X_C)^2} \angle \tan^{-1}\left(\frac{X_L - X_C}{R}\right)$$

The larger reactance in the term $(X_L - X_C)$ determines whether the circuit is overall capacitive or inductive. The current will lead or lag the source voltage depending on whether the capacitive or inductive reactance predominates. The impedance diagram and phasor diagram shown in Figures 4.1 and 4.2 are for an *R-L-C* circuit where the capacitive reactance predominates, causing a leading current. If the frequency of the source is increased sufficiently, the circuit will become overall inductive since $X_L = 2\pi f L$ increases and $X_C = 1/(2\pi f C)$ decreases.

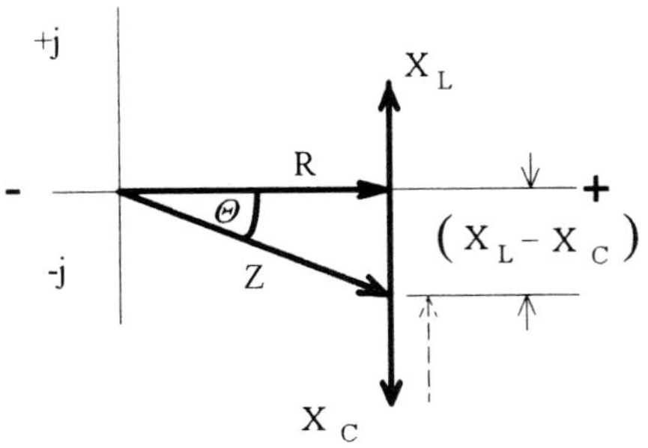

Figure 4.1

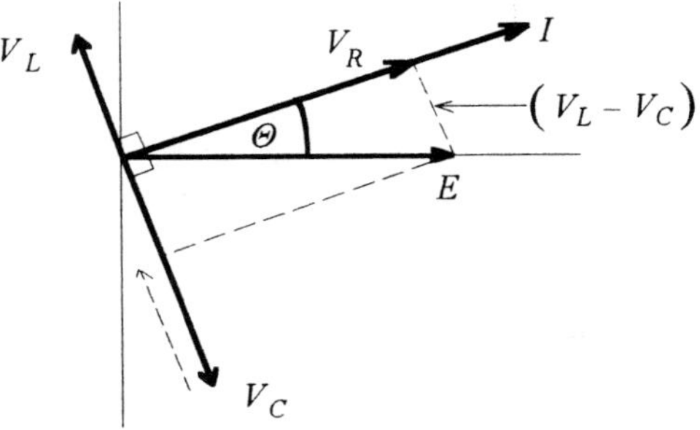

Figure 4.2

PROCEDURE Part 1

The impedance of an *R-L-C* circuit can be obtained from the solution results of the circuit provided by *Breadboard*. Using *Breadboard*, draw and solve the circuit shown in Figure 4.3 for a frequency of 20 kHz.

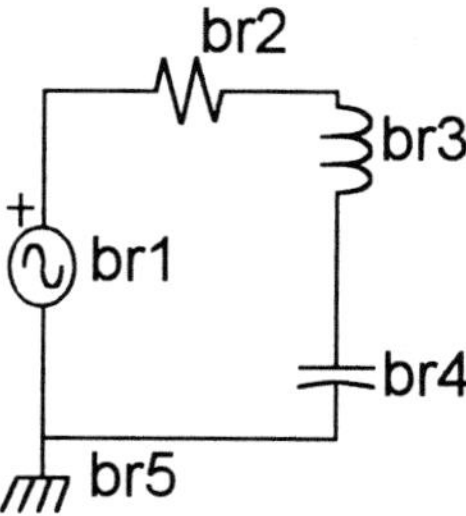

Net List for the Circuit

Br. 1: $E = 28$ V/0°
Br. 2: $R = 100\ \Omega$
Br. 3: $L = 2.0$ mH
Br. 4: $C = 0.05\ \mu F$
Br. 5: voltage reference

Figure 4.3

Obtain a printout of the net list, the current in the circuit, and the branch voltages.

From the solution results, calculate the reactances X_L and X_C and the impedance Z_T of the circuit.

$$X_L = V_L/I = \underline{\hspace{2cm}} \angle \underline{\hspace{2cm}}°\ \Omega$$

$$X_C = V_C/I = \underline{\hspace{2cm}} \angle \underline{\hspace{2cm}}°\ \Omega$$

$$Z_T = E/I = \underline{\hspace{2cm}} \angle \underline{\hspace{2cm}}°\ \Omega$$

On the printout page, draw the impedance diagram approximately to scale.

On the printout page, draw the phasor diagram for the current and the voltages of the circuit.

Show diagrammatically that the phasor sum of the voltages is $V_R + V_L + V_C$ equals the source voltage E.

What is the phase angle of the current? __________ °

Is the impedance of the circuit overall capacitive? _______

PROCEDURE Part 2

If the frequency of the source is decreased sufficiently, the capacitive reactance will dominate the X_L-X_C term in the impedance equation and the circuit will become overall capacitive. Solve the circuit for a frequency of 10 kHz.

Obtain a printout of the net list, the current in the circuit, and the branch voltages.

From the solution results, calculate the reactances X_L and X_C and the impedance Z_T of the circuit.

$$X_L = V_L/I = \underline{\hspace{2cm}} \angle \underline{\hspace{2cm}}^\circ \; \Omega$$

$$X_C = V_C/I = \underline{\hspace{2cm}} \angle \underline{\hspace{2cm}}^\circ \; \Omega$$

$$Z_T = E/I = \underline{\hspace{2cm}} \angle \underline{\hspace{2cm}}^\circ \; \Omega$$

On the printout page, draw the impedance diagram approximately to scale.

On the printout page draw the phasor diagram for the current and the voltages of the circuit.

Show diagrammatically that the phasor sum of the voltages is $V_R + V_L + V_C$ equals the source voltage E.

What is the phase angle of the current?_________$^\circ$

Is the circuit overall capacitive? _______

Write your comments below on the comparison of the solution results at the two frequencies with reference to the impedance of the circuit, the phase angle of the current, and the magnitudes of the voltages across each component.

Comments

Note: The solution results of the current through the inductor may appear to be 180° out of phase with the current in the resistor. However, the node-from and node-to information in the square brackets will indicate that the nodes have been reversed by the program.

PROCEDURE Part 3

A special case of the series *R-L-C* circuit occurs when the frequency is such that the inductive reactance equals the capacitive reactance. This condition is called resonance. At this frequency the impedance just equals the resistance alone since the reactances cancel each other and the term $(X_L - X_C)$ equals zero. At resonance the current is calculated as *E/R* and is in phase with the applied voltage. The voltages across the capacitor and the inductor are equal and can be much larger than the source voltage. The resonant frequency is calculated from the resonance condition.

$$X_L = X_C$$

$$2\pi f L = \frac{1}{2\pi f C}$$

$$fr = \frac{1}{2\pi\sqrt{LC}}$$

Solve the circuit at the resonant frequency of 15,914.94 Hz.

The phase angle of the current is ________________ mA.

The impedance Z = *E/I* = ________________ Ω.

The voltage across the capacitor = ___________ V.

The voltage across the inductor = ___________ V.

The voltage across the resistor = ___________ V.

Draw the impedance diagram and the phasor diagram approximately to scale below.

Impedance Diagram   Phasor Diagram

Parallel *R*-*L*-*C* Circuits

OBJECTIVE

To analyze the currents in a parallel *R-L-C* circuit, and to determine the admittance and impedance of the circuit.

THEORY

In the circuit of Figure 5.1 the applied voltage is common to all the components. The current in the resistor is in phase with the applied voltage. The currents in the inductive and capacitive branches respectively lag and lead the voltage by 90°. To obtain the total current drawn from the source, the currents must be added in phasor form.

The admittance of the circuit in conductance and susceptance form is

$$Y = G + j(B_C - B_L) \quad \text{Siemens}$$

where $G = 1/R$, $\quad B_C = 1/X_C$, and $B_L = 1/X_L$.

The admittance of the circuit can be obtained from the solution as $Y_T = I_T/E$.

PROCEDURE Part 1

Using *Breadboard* draw and solve the circuit in Figure 5.1 for a frequency of 25 kHz.

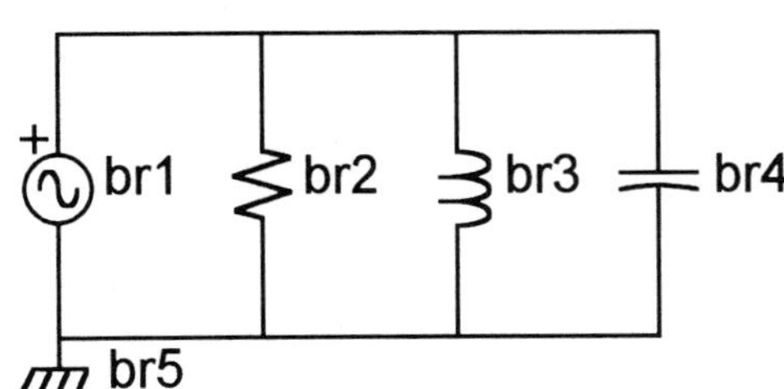

Net List for the Circuit

Br. 1: E = 100 V/0°
Br. 2: R = 100 Ω
Br. 3: L = 1.8 mH
Br. 4: C = 0.1 μF
Br. 5: voltage reference

Figure 5.1

Obtain a printout of the schematic, net list, branch voltages, and branch currents for the circuit.

Mark the current in each branch on the schematic.

From the solution results, calculate the value of:

$$G = I_R/V_R = \underline{\hspace{3cm}} \text{ mS}$$

$$B_L = I_L/V_L = \underline{\hspace{3cm}} \text{ mS}$$

$$B_C = I_C/V_C = \underline{\hspace{3cm}} \text{ mS}$$

Calculate the admittance as $Y_T = G + j(B_C - B_L) = \underline{\hspace{3cm}} \angle \underline{\hspace{2cm}}° \text{ mS}$

and $Y_T = I_T/E = \underline{\hspace{3cm}} \angle \underline{\hspace{2cm}}° \text{ mS}$.

On the printout page, draw both the phasor diagram for E, I_R, I_C, I_L and the admittance diagram.

Prove by phasor addition:

$$I_T = I_R + I_C + I_L = \underline{\hspace{3cm}} \angle \underline{\hspace{2cm}}° \text{ A}$$

PROCEDURE Part 2

If the frequency of the source is changed, the reactance of the capacitor and the inductor will also change. The overall impedance of the circuit may become leading or lagging. The current drawn from the source will change and may be either leading or lagging. Repeat the above procedure for a frequency of 10 kHz.

Obtain a printout of the schematic, net list, branch voltages, and branch currents for the circuit. On the schematic, mark the current in each branch.

From the solution results, calculate the following values:

$$G = I_R/V_R = \underline{\hspace{3cm}} \text{ mS}$$

$$B_L = I_L/V_L = \underline{\hspace{3cm}} \text{ mS}$$

$$B_C = I_C/V_C = \underline{\hspace{3cm}} \text{ mS}$$

Calculate the admittance as $Y_T = G + j(B_C - B_L) = \underline{\hspace{2cm}} \angle \underline{\hspace{1.5cm}}° \text{ mS}$

and $Y_T = I_T/E = \underline{\hspace{2cm}} \angle \underline{\hspace{1.5cm}}° \text{ mS}$

On the printout page draw the phasor diagram for E, I_R, I_C, I_L and the admittance diagram.

Prove by phasor addition $I_T = I_R + I_C + I_L = \underline{\hspace{3cm}} \angle \underline{\hspace{2cm}}° \text{ A}$

FURTHER ANALYSIS

Choose a problem from your text on *R-L-C* parallel circuits. Solve the circuit by the method of admittances. Then use *Breadboard* to check your results.

OBJECTIVE

To obtain the solution of AC series-parallel circuits and to analyze the results.

THEORY

Network reduction is the usual method of solving for the circuit parameters of single-source AC circuits. The process involves combining parallel branches to obtain the equivalent impedance and adding the series impedances until the network has been reduced to a single equivalent impedance. The total current can then be found and by applying the current divider rule the current in each branch can then be obtained. All the calculations must be performed using complex algebra.

Breadboard readily provides the solution for the currents and voltages of each branch. It is not sufficient for a student just to obtain the solutions. The solution results need to be analyzed to obtain an understanding of AC circuit behavior.

PROCEDURE Part 1

Draw and solve the circuit of Figure 6.1 for a frequency of 60 Hz.

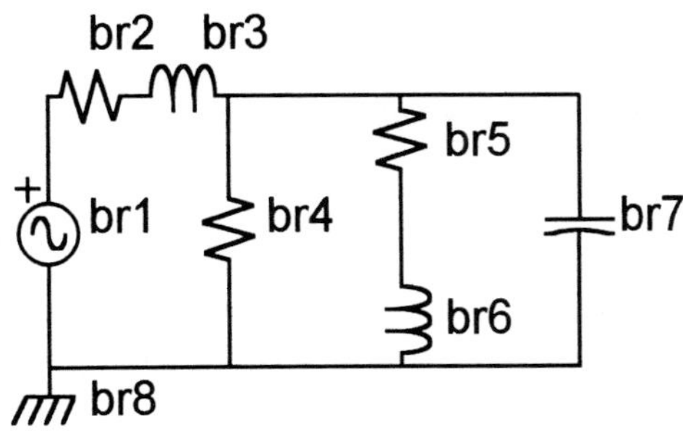

Figure 6.1

Net List for the Circuit

Br. 1: E = 240 V/0°
Br. 2: R = 80 Ω
Br. 3: L = 0.2 H
Br. 4: R = 200 Ω
Br. 5: R = 400 Ω
Br. 6: L = 1.8 H
Br. 7: C = 10 μF
Br. 8: voltage reference

Obtain a printout of the schematic diagram, the net list, the branch currents, and the branch voltages.

Record on the circuit diagram the current and voltage (magnitude and phase angle) of each branch.

Calculate the impedance of the circuit.

$$Z_T = E/I_T = \underline{\hspace{2cm}} \angle \underline{\hspace{2cm}}° \; \Omega$$

Confirm Kirchhoff's current law by summing the current in branches 4, 5, and 7, and compare to the current in branch 3. Phasor addition must be used.

$$I_4 + I_5 + I_7 = \underline{\hspace{2cm}} \angle \underline{\hspace{2cm}}° \; A$$

Current in branch 3 = $\underline{\hspace{2cm}} \angle \underline{\hspace{2cm}}°. \; A$

PROCEDURE Part 2

Draw and solve the circuit shown in Figure 6.2 for a frequency of 1.0 kHz. Analyze the circuit.

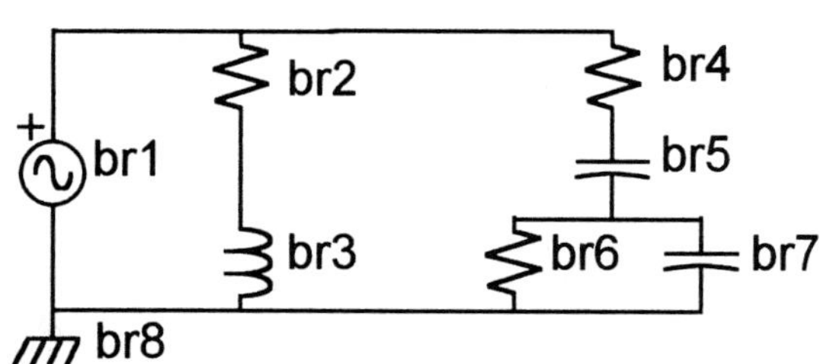

Figure 6.2

Net List for the Circuit

Br. 1: E = 40 V/0°
Br. 2: R = 1.2 kΩ
Br. 3: L = 10 mH
Br. 4: R = 1 kΩ
Br. 5: C = 0.1 µF
Br. 6: R = 2.7 kΩ
Br. 7: C = 0.05 µF
Br. 8: voltage reference

Obtain a printout of the schematic diagram, the net list, the branch currents, and the branch voltages.

On the circuit diagram, record the currents and voltages (magnitude and phase angle).

Calculate the impedance of the circuit.

$$Z_T = E/I_T = \underline{\hspace{2cm}} \angle \underline{\hspace{2cm}}° \; \Omega$$

Sum the currents in branches 2 and 4, using phasor addition and compare to the current in branch 1.

$$I_2 + I_4 = \underline{\hspace{2cm}} \angle \underline{\hspace{2cm}}° \; A$$

Current in branch 1 = $\underline{\hspace{2cm}} \angle \underline{\hspace{2cm}}° \; A$

OBJECTIVE

To study the frequency response characteristics of *R-C* and *R-L* circuits with application to high-pass and low-pass filters.

THEORY

Filter circuits take advantage of the frequency characteristics of the reactance of capacitors and inductors to block or pass signals within a desired frequency range. An example is the speaker crossover network in audio systems.

The low pass filter of Figure 7.1 may be considered a voltage divider where the output voltage is taken across the capacitor. As the frequency of the source is increased to a higher value, the reactance of the capacitor will decrease and the output voltage will decrease. The graph of Figure 7.2 shows the output voltage as a function of frequency for a **low-pass** *R-C* filter.

Filter networks are described by their cutoff or corner frequency. The cutoff frequency is the frequency at which the output voltage has dropped to 0.707 of the maximum output voltage. The cut-off frequency is also referred to as the half-power or the 3 db down frequency. At this frequency the capacitive reactance equals the resistance of the circuit.

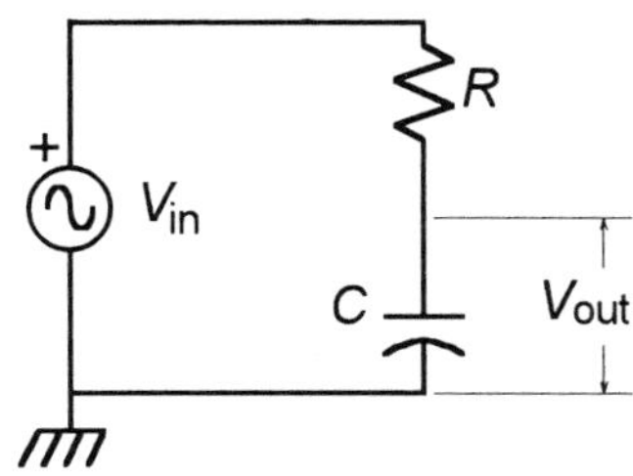

Figure 7.1

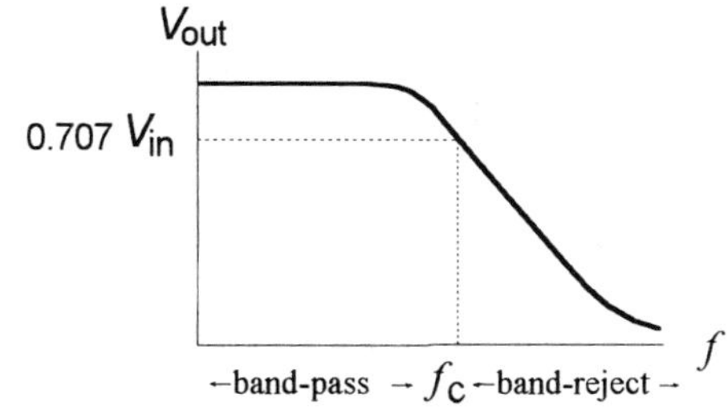

Figure 7.2

For the **low-pass** *R-C* filter the phase angle of the **voltage output** will vary from -90° at a very low frequency to 0° at high frequencies. An application of this characteristic is in lag circuits where phase relationship is significant.

The voltage output of the *R-C* circuit may be calculated as shown below.

by the potential divider rule $\quad V_O = \dfrac{X_C}{\sqrt{R^2 + X_C{}^2}} V_{\text{IN}}$

at $f_C \quad R = X_C = \dfrac{1}{2\pi f C} \qquad$ and $\quad f_C = \dfrac{1}{2\pi R C}$

PROCEDURE Part 1

The circuit of Figure 7.1 is a low-pass filter with branch 4 the output load resistor across which the output voltage of the circuit is to be measured.

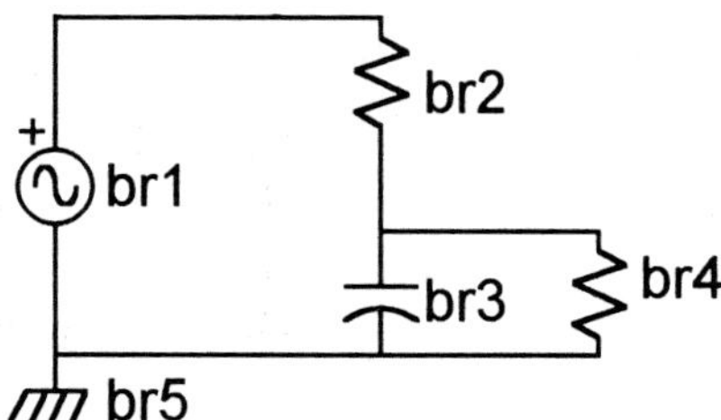

Net List for the Circuit

Br. 1: E = 10 V/0°
Br. 2: R = 1.5915 kΩ
Br. 3: C = 0.1 μF
Br. 4: R = 1.0 MΩ
Br. 5: voltage reference

Figure 7.3

Draw the circuit for the values shown in the net list and obtain a graph of the voltage output (node 2) over the range of frequencies from 100 Hz to 100 kHz. A tabular listing of the data for the graph will facilitate obtaining the corner frequencies.

Calculate the corner frequency from the formula given above.

$$f_C = \underline{\hspace{3cm}} \text{ Hz}$$

From the graph, obtain the corner frequency ($V_o = 0.707$ V).

$$f_C = \underline{\hspace{3cm}} \text{ Hz}$$

Calculate the phase angle of the voltage output at the corner frequency.

$$\theta = \tan^{-1}(X_C/R) - 90°$$

From the graph determine the phase angle of the voltage output at the corner frequency.

phase angle at $f_C = \theta \underline{\hspace{2.5cm}}$ °

PROCEDURE Part 2

A low-pass filter can also be obtained with an *R-L* circuit as shown in Figure 7.4. As the frequency is increased in the *R-L* low-pass filter circuit, the voltage drop across the inductor increases since $X_L = 2\pi f L$ and the voltage output across branch 4 will decrease. The voltage output is given by the formula

$$\text{by the potential divider rule } V_O = \frac{R}{\sqrt{R^2 + X_L^2}} V_{IN}$$

$$\text{at } f_C \quad R = X_L = 2\pi f L \quad \text{and} \quad f_C = \frac{R}{2\pi L}$$

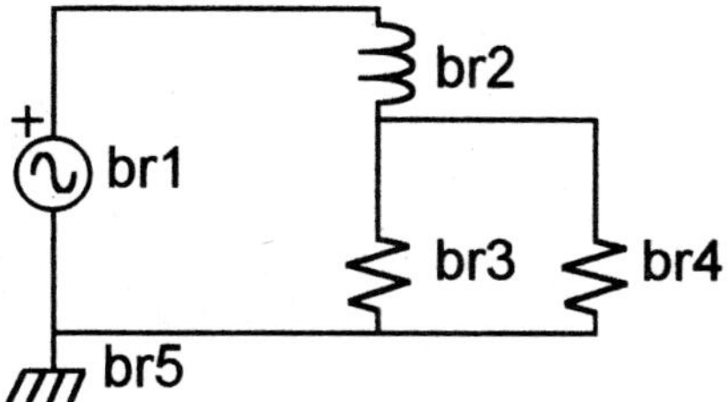

Figure 7.4

Net List for the Circuit

Br. 1: E = 10 V/0°
Br. 2: R = 1.5914 kΩ
Br. 3: L = 150 mH
Br. 4: R = 1.0 MΩ
Br. 5: voltage reference

Draw the *R-L* circuit for the values shown in the net list and obtain a graph of the voltage output (node 2) over the range of frequencies from 100 Hz to 100 kHz.

Calculate the corner frequency from the formula given above.

$$f_C = \underline{\hspace{3cm}} \text{ Hz}$$

From the graph, obtain the corner frequency ($V_o = 0.707$ V).

$$f_C = \underline{\hspace{3cm}} \text{ Hz}$$

Calculate the phase angle of the voltage output at the corner frequency.

$$\theta = 90° - \tan^{-1}(X_L/R)$$

From the graph, determine the phase angle of the voltage output at the corner frequency.

$$\text{phase angle at } f_C = \theta = \underline{\hspace{3cm}} °$$

PROCEDURE Part 3

Bode plots are graphs of the voltage output in decibels versus frequency of filter circuits on a logarithmic scale. The definition of a decibel is given below.

$$dB = 10 \log(P_o/P_{in})$$

If the power input and power output are taken across identical impedances, the formula can be modified as a ratio of the input and output voltages

$$dB = 10 \log(V_o^2/V_{in}^2) = 20 \log(V_o/V_{in}).$$

The Bode diagram is actually a straight line approximation of the frequency response. The corner frequency is calculated and plotted and the roll-off can then be drawn from this point at the rate of 10 dB per decade for the simple *R-L* or *R-C* circuit.

Breadboard will provide graphs plotted on the basis of decibels versus frequency.

For the low-pass filter shown in Figure 3, obtain a Bode plot over the frequency range of 100 Hz to 100 kHz. Draw straight line approximations as in a Bode plot and determine the corner frequency.

The corner frequency = ________________ Hz.

Compare your result to that obtained in procedure 1.

FURTHER ANALYSIS

Repeat the above analysis for a high-pass *C-R* filter and a high-pass *R-L* filter for the same value of components as in the circuits above, and over the same frequency range. See Figures 7.5 and 7.6.

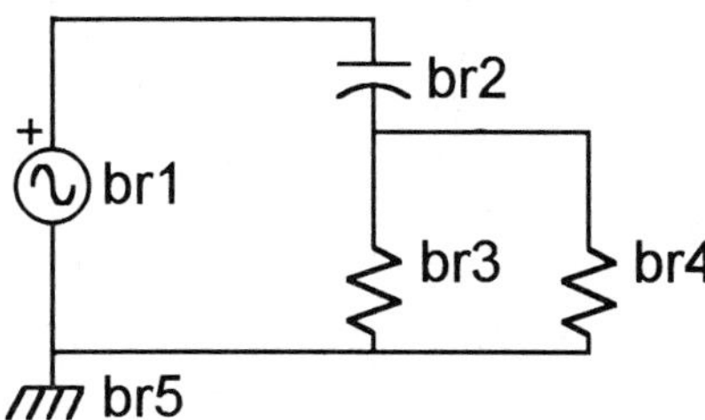

Figure 7.5

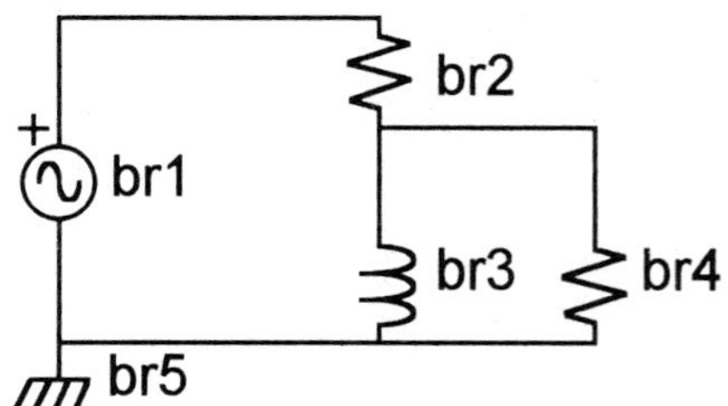

Figure 7.6

Circuits with More than One Source

AC Exercise

OBJECTIVE

To obtain the solutions and to analyze circuits containing more than one source.

THEORY

Networks with more than one source would require the methods of mesh or nodal analysis to effect a solution. Circuit-analysis programs, such as *Breadboard,* can provide the solutions readily.

PROCEDURE Part 1

The circuit of Figure 8.1 represents two alternators supplying power to a common load. The generated voltage of each source has been set at $120\angle 0°$ but the alternators have different internal impedances. Perform an analysis of the circuit to determine the load current and power provided by each alternator.

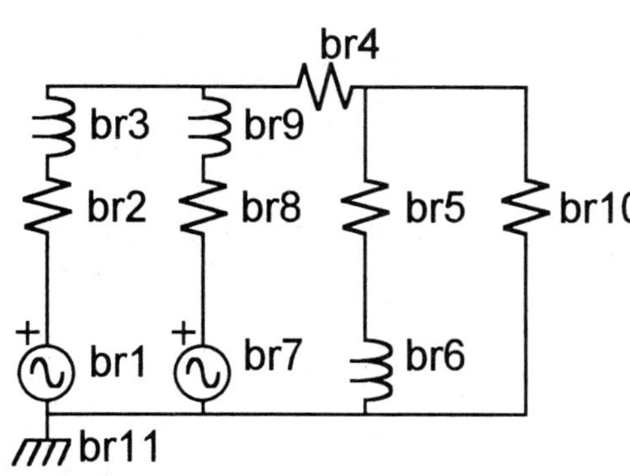

Figure 7.1

Net List for the Circuit

Br. 1: E = 120 V/0°
Br. 2: R = 3 Ω
Br. 3: L = 8 mH
Br. 4: R = 0.02 Ω
Br. 5: R = 12 Ω
Br. 6: L = 18 mH
Br. 7: E = 120 V/0°
Br. 8: R = 4 Ω
Br. 9: L = 10 mH
Br. 10: R = 10 Ω
Br. 11: voltage reference

Draw and solve the circuit for a frequency of 60 Hz.

The current delivered by the alternator in branch 1 = _______________ A.

The current provided by the alternator in branch 7 = _______________ A.

The power delivered by the alternator in branch 1 = _______________ W.

The power delivered by the alternator in branch 7 = _______________ W.

79

From the solution results for the currents in the branches, draw the phasor diagram for the currents delivered by the generators and the total load current (**I**, branch 4).

Phasor Diagram

PROCEDURE Part 2

The circuit in Figure 8.2 simulates a two-phase, 400 Hz control system. Perform an analysis of the currents in each branch.

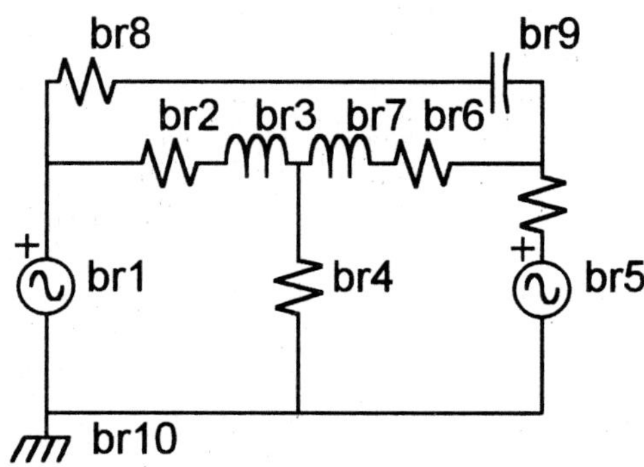

Figure 8.2

Net List for the Circuit

Br. 1: $E = 60$ V/0°
Br. 2: $R = 4\ \Omega$
Br. 3: $L = 2$ mH
Br. 4: $R = 180$
Br. 5: $E = 60$ V/90°
Br. 6: $R = 4\ \Omega$
Br. 7: $L = 2$ mH
Br. 8: $R = 78\ \Omega$
Br. 9: $C = 4\ \mu F$
Br. 10: voltage reference

Draw and solve the circuit for the current in each branch.

Record the voltages and currents on the circuit diagram.

FURTHER ANALYSIS

Repeat the above procedure for the circuit of Figure 8.1 if the voltage of the alternator in branch 7 is 120 V/10°.

Repeat the procedure if the voltage of the alternator is 125 V/0°.

Power in AC Circuits

OBJECTIVE

To determine the power, reactive power, and apparent power of an AC circuit, and to draw the power triangle.

THEORY

Power is the rate of flow of energy (Joules per second or Watts). In AC networks power is developed only in the resistive elements. The total power dissipated is the sum of the individual branch powers whether they are series or parallel connected.

$$\text{Power (P)} = I^2R = V_R I = V_R^2/R = V_R I \cos\theta$$

Reactive power or quadrature power (Q) is the measure of the rate of energy exchanged (not absorbed) by the reactive elements of the circuit and the source. The units of reactive power is reactive-volt-amperes (vars) and may be capacitive or inductive. The total reactive power of a network which has to be provided by the source is the difference of the sum of the capacitive vars and the inductive vars.

$$\text{Reactive power (Q)} = I^2X = V_X I = V_X^2/X = V_X I \sin\theta$$

Apparent power (S) is the product of voltage and current of a branch of the network. The total apparent power is the product of the source voltage and current.

$$\text{Apparent power (S)} = VI = I^2Z = \sqrt{(P^2 + Q^2)}$$

Power factor (*Fp*) is defined as the ratio of the real power to the apparent power and is described as either leading (capacitive) or lagging (inductive).

$$\text{Power factor (Fp)} = P/S = \cos\theta$$

The total P, Q, S delivered by the source is

$$P_T = EI_T \cos\theta \qquad Q_T = EI_T \sin\theta \qquad S_T = EI_T$$

A power triangle may be constructed to show the relationship of P, Q, and S, similar to an impedance diagram. Figure 9.1 is the power triangle for an inductive load and a capacitive load.

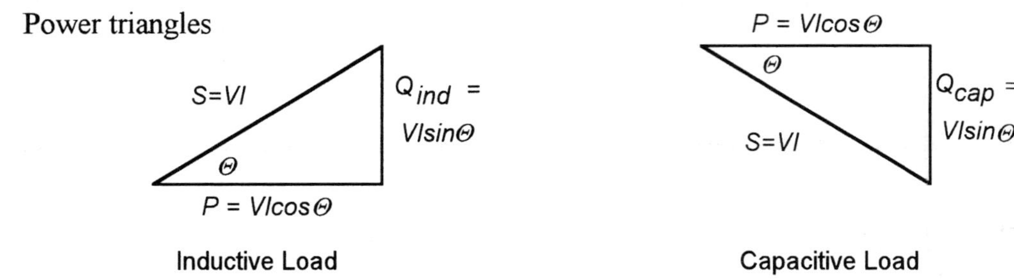

Power triangles

Figure 9.1

PROCEDURE

The circuit shown in Figure 9.2 represents an AC source providing power to an induction motor load, and a synchronous motor load in an industrial plant.

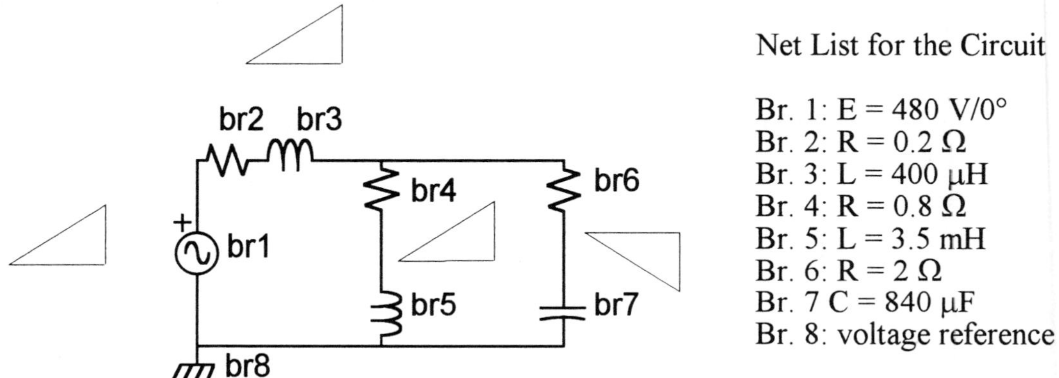

Net List for the Circuit

Br. 1: $E = 480$ V/0°
Br. 2: $R = 0.2\ \Omega$
Br. 3: $L = 400\ \mu H$
Br. 4: $R = 0.8\ \Omega$
Br. 5: $L = 3.5$ mH
Br. 6: $R = 2\ \Omega$
Br. 7 $C = 840\ \mu F$
Br. 8: voltage reference

Figure 9.2

Draw and solve the circuit of Figure 9.2 for a frequency of 60 Hz. Mark on the power triangles the watts, vars and volt-amperes for each branch.

The sum of the power in each branch $= P_{\text{branch 2}} + P_{\text{branch 4}} + P_{\text{branch 6}} =$ _______ W

equals the power of the source $(P_{\text{branch 1}}) =$ _______________ W.

The sum of the reactive power $= Q_{\text{branch 3}} + Q_{\text{branch 5}} - Q_{\text{branch 7}} =$ _____________ vars

equals the vars from the source $(Q_{\text{branch 1}}) =$ _____________ vars.

The overall power factor (Fp) for the circuit is $P_T/S_T =$ _________ (lead/lag).

Calculate the total power from the relationship $P_T = EI\cos\theta = EI(Fp)$ _______ W.

Calculate the power factor from the phase angle between the source voltage and current. $Fp = \cos\theta =$ _____________

Power Factor Correction 10

OBJECTIVE

To observe the effect of adding capacitors to an inductive load on the power factor of the network and on the magnitude of the source current.

THEORY

Industrial loads, which are generally inductive, cause a larger current to be drawn from the power source than would be necessary if the load were purely resistive. The source must provide the reactive power required by the load in addition to the real power.

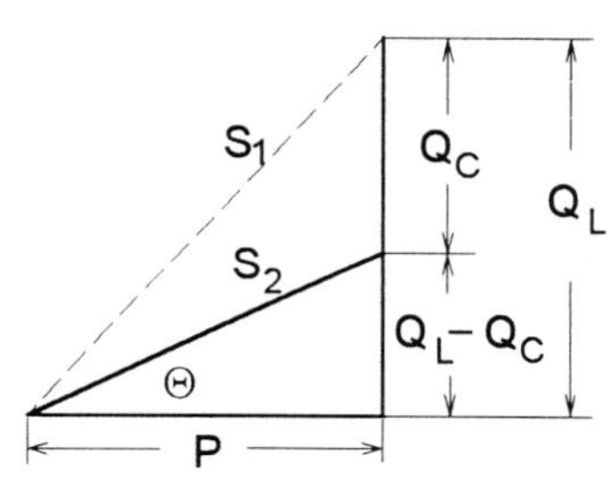

Figure 10.1

This necessitates larger conductors and transformers to carry the heavier currents, with subsequent higher costs and power losses. By adding capacitors to the plant load to compensate for the inductive vars (Q_L) of the load a smaller total current is drawn from the lines. The apparent power (S_T) is reduced and approaches the value of the power (P_T). The power-factor angle is also reduced such that the power factor ($Fp = \cos\theta$) approaches unity. The power triangle in Figure 10.1 shows this effect diagrammatically.

PROCEDURE Part 1

The circuit of Figure 10.2 represents an industrial load supplied by a source. The resistors in branches 2 and 4 represent the resistance of the power-distribution lines.

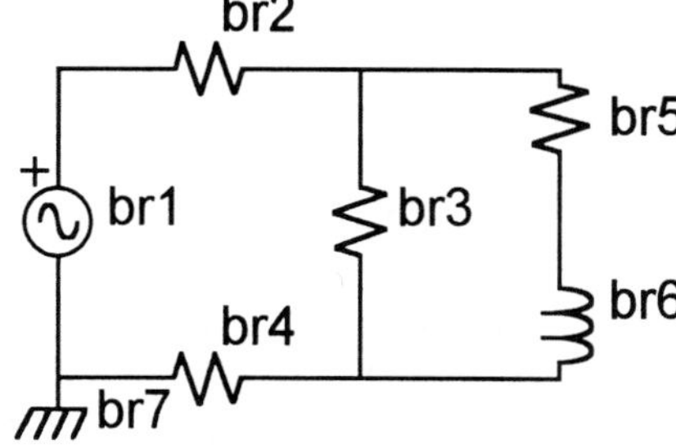

Figure 10.2

Net List for the Circuit

Br. 1: E = 480 V/0°
Br. 2: R = 0.06 Ω
Br. 3: R = 12 Ω
Br. 4: R = 0.06 Ω
Br. 5: R = 5 Ω
Br. 6: L = 25 mH
Br. 7: voltage reference

Draw and solve the circuit for a frequency of 60 Hz.

Draw and label the total power triangle and the phasor diagram for E and I approximately to scale.

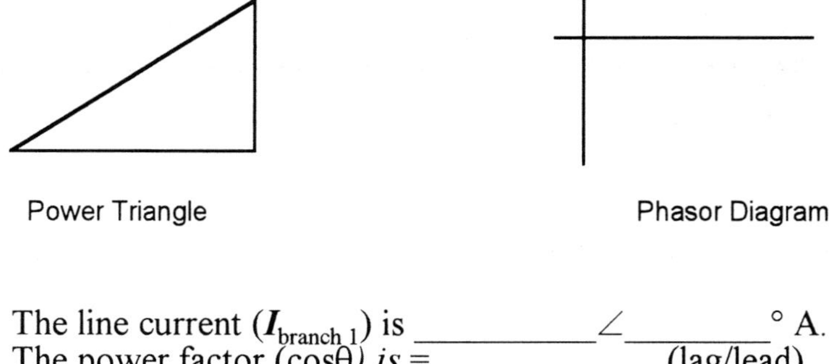

Power Triangle   Phasor Diagram

The line current ($I_{branch\ 1}$) is ___________ $\angle$ _________° A.
The power factor ($\cos\theta$) *is* = __________ (lag/lead).
The line losses are ($P_{branch\ 2}$ + $P_{branch\ 4}$) __________ W.

PROCEDURE Part 2

Add a 100 µF capacitor (branch 8) in parallel with the load as in Figure 10.3.

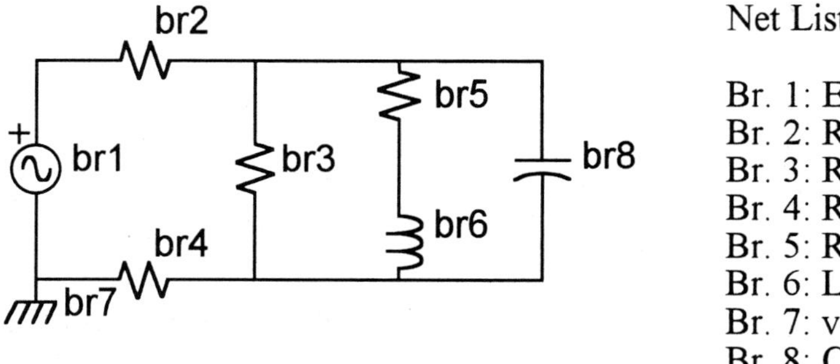

Net List for the Circuit

Br. 1: E = 480 V/0°
Br. 2: R = 0.06 Ω
Br. 3: R = 12 Ω
Br. 4: R = 0.06 Ω
Br. 5: R = 5 Ω
Br. 6: L = 25 mH
Br. 7: voltage reference
Br. 8: C = 100 µF

Figure 10.3

Solve the circuit for a frequency of 60 Hz.

Draw and label the total power triangle and the phasor diagram for E and I approximately to scale.

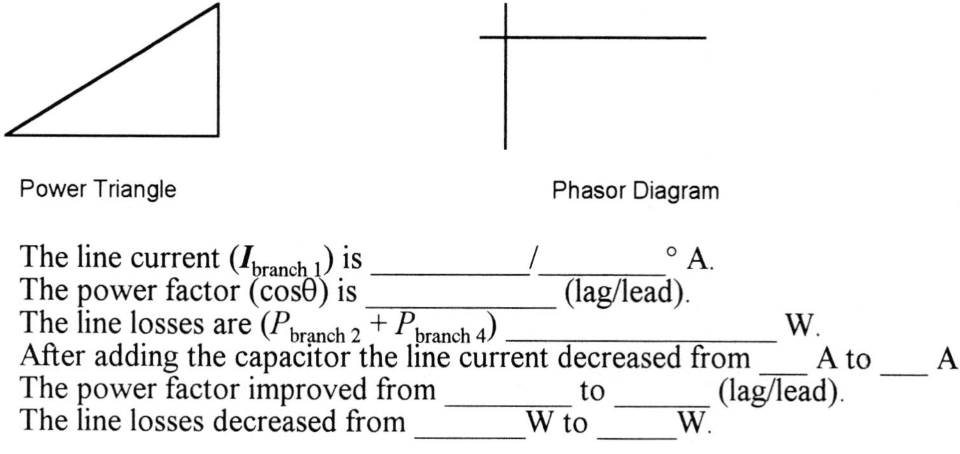

Power Triangle   Phasor Diagram

The line current ($I_{branch\ 1}$) is ___________ / _________° A.
The power factor ($\cos\theta$) is __________ (lag/lead).
The line losses are ($P_{branch\ 2}$ + $P_{branch\ 4}$) __________ W.
After adding the capacitor the line current decreased from ___ A to ___ A.
The power factor improved from ________ to ______ (lag/lead).
The line losses decreased from ________ W to ______ W.

Series Resonance

OBJECTIVE

To study the frequency response of a series resonant circuit.

THEORY

The impedance of a series R-L-C circuit is given as $\mathbf{Z_T} = \mathbf{R} + j(X_L - X_C)$. At one specific frequency, called the resonant frequency, the capacitive reactance equals the inductive reactance and the impedance of the circuit is just the resistance alone.

At resonance $X_L = X_C$ and $2\pi f_r L = 1/(2\pi f_r C)$

$$\text{from which} \quad f_r = \frac{1}{2\pi\sqrt{LC}} \text{ Hz}$$

The variation of current as the frequency is varied is shown in Figure 11.1. At the resonant frequency the impedance of the circuit will be a minimum and the current will be maximum and in phase with the source voltage.

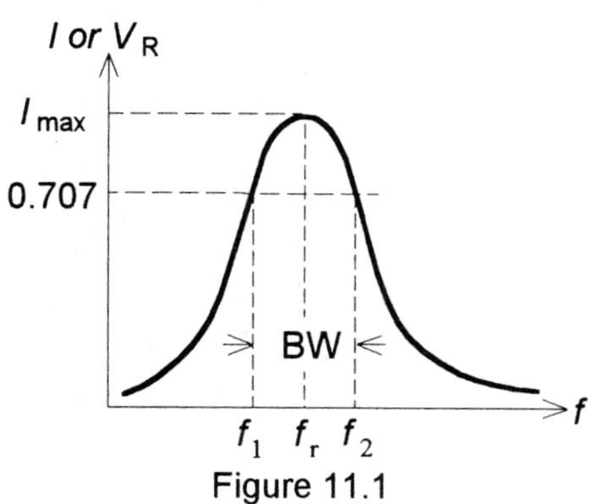

Figure 11.1

The characteristic resonant rise of current in the circuit, and therefore the voltage across the components, is made use of in frequency selective circuits such as the tuning circuit in radio receivers. The width of the curve that determines the selectivity $(f_2 - f_1)$ decreases as the resistance decreases and as the ratio of L to C increases. A quality factor describing the selectivity of the circuit is defined as the Q of the circuit.

By definition $\quad Q = \dfrac{X_L}{R} = \dfrac{2\pi f_r L}{R}$ and since $f_r = \dfrac{1}{2\pi\sqrt{LC}} \quad Q = \dfrac{1}{R}\sqrt{\dfrac{L}{C}}$

The higher the value of Q the more selective the circuit, and values of Q from 10 to 200 are common. A practical coil has some resistance due to the resistance of the wire conductor making up the coil. In the analysis above this resistance has been included in the total resistance of 100 Ω.

The series resonant circuit is used to pass or reject a band of frequencies, and the bandwidth of the circuit is defined as $BW = f_2 - f_1$ where the upper and lower cutoff frequencies are the half power levels or 0.707 of the maximum. For high Q circuits ($Q \geq 10$) the bandwidth may be considered as centered about the resonant frequency and is given as the formula below. For low Q circuits ($Q \leq 10$) the bandwidth is not centered about the resonant frequency.

$$BW = f_2 - f_1 = \frac{f_r}{Q} = \frac{R}{2\pi L}$$

At the resonant frequency the voltage across the inductor and capacitor are equal and the magnitude is Q times as large as the source voltage.

$$V_L = V_C = IX_L = \frac{E}{R}X_L = QE$$

The voltage output of the circuit is taken across the capacitor.

PROCEDURE Part 1

For the component values shown in the net list of 11.2 calculate:

$f_r = $ _______________________ Hz
$Q = $ _______________________
$X_L = $ _______________________ Ω
$BW = $ _______________________ Hz

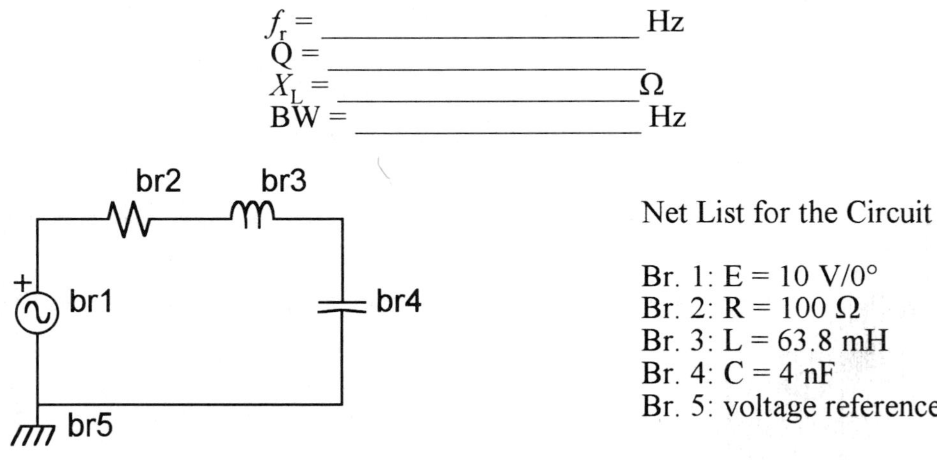

Figure 11.2

Draw and solve the circuit of Figure 11.2 for the resonant frequency calculated above. On the circuit diagram record the current and the voltages across each component.

Calculate V_C as $QE = $ ________ V.   The solution value of $V_C = $ ______ V.

Confirm by calculation that the current in the circuit is E/R ________ $\angle$ ____ ° mA.

The calculated current I is $E/R = $ ________ $\angle$ ______ °.

PROCEDURE Part 2

The circuit of Figure 11.2 is a **high Q** circuit and has a narrow bandwidth which can be calculated as BW = f_r/Q. Obtain a graph of the voltage output across the capacitor (node 3) over the frequency range of 6 kHz to 14 kHz. If a printer is available, a hard copy of the graph will facilitate the determination of the bandwidth.

The maximum voltage across the capacitor is V_C = ________________ V.

The Q of the circuit is Q = V_C/E = ________________ .

Determine the upper and lower bandwidth frequencies that occur at the voltage level where V_C is 0.707 of the maximum V_C.

f_1 = ________________ Hz f_2 = ________________ Hz

The bandwidth = BW = f_2-f_1 = ________________ Hz.

The bandwidth can also be calculated as BW = f_r/Q = ________________ Hz.

What is the phase angle of the voltage V_C at the cutoff frequencies?

phase angle at f_1 = ________________ °
phase angle at f_2 = ________________ °

PROCEDURE Part 3

To investigate the characteristics of a **low Q** series resonant circuit change the ratio of L to C as in the circuit shown in Figure 11.3. For the component values shown in the net list calculate:

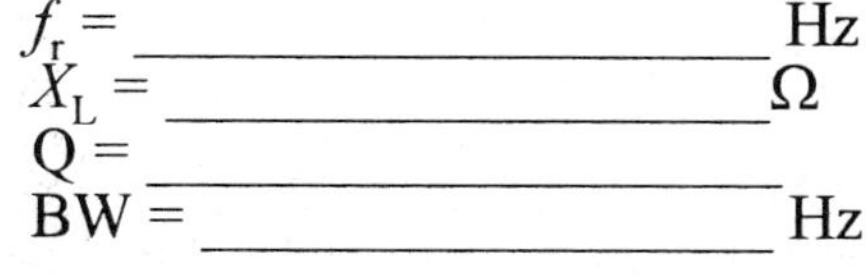

f_r = ________________ Hz
X_L = ________________ Ω
Q = ________________
BW = ________________ Hz

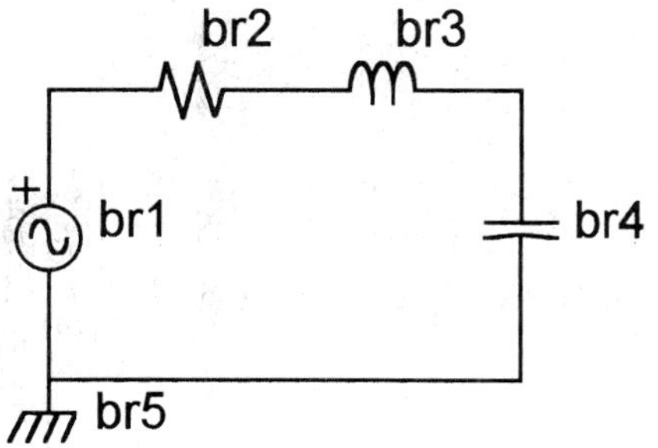

Figure 11.3

Net List for the Circuit

Br. 1: E = 10 V/0°
Br. 2: R = 100 Ω
Br. 3: L = 6.38 mH
Br. 4: C = 0.04 μF
Br. 5: voltage reference

Draw and solve the circuit of Figure 11.3 for the resonant frequency calculated above. On the circuit diagram record the current and the voltages across each component.

Calculate V_C as $QE =$ _______ V.

The solution value of V_C ($V_{\text{branch 4}}$) = _______ V.

Confirm by calculation that the current in the circuit is E/R.

The calculated current I is $E/R =$ _______ $\angle$ _______ ° mA.

The solution value of the current in the circuit is _______ $\angle$ _______ mA.

The low Q circuit of Figure 11.3 has a wide bandwidth. Obtain a graph of the voltage output across the capacitor (node 3) over the frequency range of 6 kHz to 14 kHz. If a printer is available, a hard copy of the graph will facilitate the determination of the bandwidth.

The maximum voltage across the capacitor is $V_C =$ _______________ V.

The Q of the circuit is Q = $(\text{max}V_C)/E =$ _________________ .

Determine the upper and lower bandwidth frequencies which occur at the voltage level where V_C is 0.707 of the maximum V_C.

$f_1 =$ _______________ Hz $f_2 =$ _____________ Hz

The bandwidth = BW = $f_2 - f_1 =$ _________________ Hz.

What is the phase angle of the voltage V_C at the cutoff frequencies?

phase angle at $f_1 =$ _____________ °
phase angle at $f_2 =$ _____________ °

FURTHER ANALYSIS

The selectivity of a series resonant circuit is also dependent on the resistance of the circuit. Decreasing the resistance of the low Q circuit in Figure 11.3 to 10 Ω will increase the Q of the circuit by a factor of 10; however, the current and the voltage across the capacitor will also change. Repeat the analysis for part 3 above for a circuit resistance of 10 Ω.

Parallel Resonance

OBJECTIVE

To study the frequency response of a parallel resonant circuit.

THEORY

The admittance of a parallel *R-L-C* circuit with **ideal** components as shown in Figure 12.1 is given as

$$Y_T = G + j(B_C - B_L) = \frac{1}{R} + j\left(\frac{1}{X_C} - \frac{1}{X_L}\right)$$

At the resonant frequency the susceptance of the inductor and the capacitor are equal. The input impedance of the circuit is a maximum and is purely resistive. These characteristics make the circuit useful as a frequency-selective circuit in electronics.

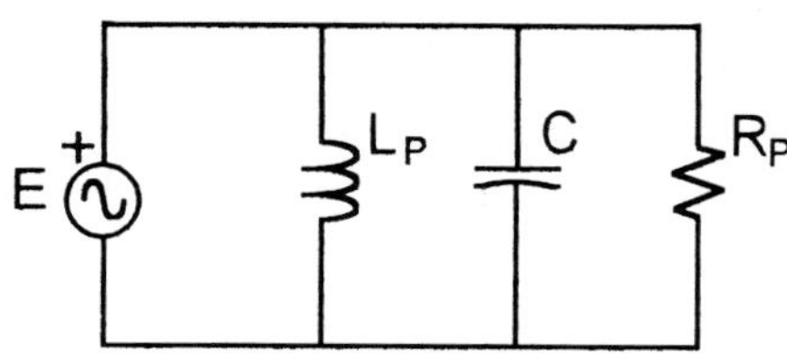

Figure 12.1

At resonance $B_L = B_C$

from which $\quad 2\pi f_p C = \dfrac{1}{2\pi f_p L}$

and the resonant frequency $\quad f_P = \dfrac{1}{2\pi\sqrt{LC}}$

At frequencies above and below the resonant frequency, the capacitive or inductive reactance in parallel with the resistance causes the impedance of the circuit to be less than that at resonance. At resonance the current drawn from the source is minimum and in phase with the voltage of the source. This characteristic is the opposite of the series resonant circuit and is shown in the graph of current versus frequency in Figure 12.2.

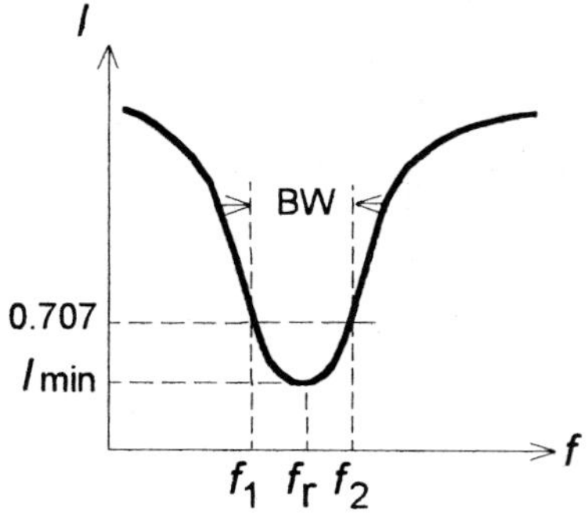

Figure 12.2

A practical parallel resonant circuit, Figure 12.3, requires that the resistance of the coil (R_l) be included in the calculation of the resonant frequency. The capacitor is assumed to be ideal with practically no leakage current. The unity power factor resonant frequency becomes

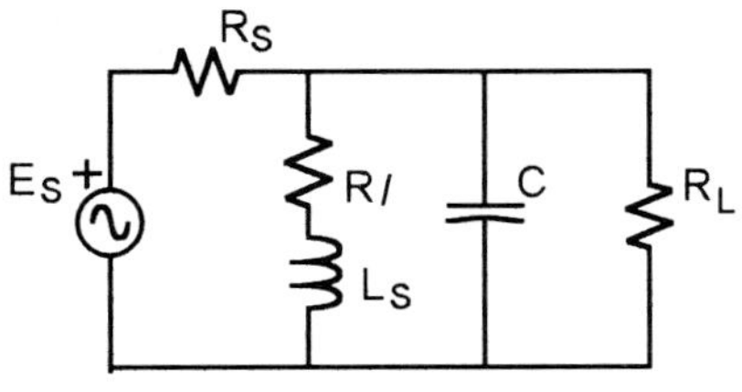

Figure 12.3

$$f_p = \frac{1}{2\pi\sqrt{LC}}\sqrt{1-\frac{R_l^2 C}{L}}$$

Note that the resonant frequency is a function of the resistance of the coil and that if R_l is large (low Q coil) resonance may not occur. If the coil is a high Q coil ($Q\geq 10$), the formula developed for the ideal circuit is a close approximation.

$$f_m = \frac{1}{2\pi\sqrt{LC}}\sqrt{1-\frac{1}{4}\left(\frac{R_l^2 C}{L}\right)}$$

The frequency at which maximum impedance occurs (f_m) is slightly different from that for unity power factor.

For a practical parallel resonant circuit the selectivity is determined by the total equivalent parallel resistance and the equivalent parallel inductance. Refer to Figure 12.1 and consider R_p to be the total equivalent parallel resistance of the practical circuit in parallel with the equivalent parallel inductance L_p of the coil and the capacitor. The Q_p of the circuit becomes

$$Q_p = \frac{R_{PT}}{X_{LP}} = R_{PT}\sqrt{\frac{C}{L_P}}$$

Note that the Q_p of the circuit is directly dependent on the parallel resistance. Since the bandwidth of the circuit is $f_2-f_1 = f_r/Q_p$, the larger the parallel resistance the larger is the Q_p for given values of L and C and the smaller is the bandwidth. The value of output load resistance connected across the parallel circuit will lower the Q_p and increase the bandwidth. Also, if the tank circuit is supplied from a current source such as a transistor, the resistance of the source (collector resistor) is in parallel with the circuit and would further reduce Q_p.

In a parallel resonant circuit the tank current (L and C in parallel) will be Q_p times as large as the current supplied from the source.

$$I_C = I_S Q_p$$

The impedance at resonance will be the parallel equivalent of the coil resistance.

$$Z_p = (1 + Q_p^2)R_l$$

PROCEDURE Part 1

The parallel resonant circuit of Figure 12.4 is a high Q circuit. For the values of components in the net list calculate:

The resonant frequency f_p = ________________ Hz.

The inductive reactance $X_L = 2\pi f L$ = ________________ Ω.

The Q of the coil = X_L/R_l = ________________ .

Since this is a high Q coil, the Q_p of the circuit will be the same as the Q of the coil.

The impedance of the tank circuit is $Z_p = (1 + Q_p{}^2)R_l$ ________________ Ω.

The bandwidth $f_2{-}f_1 = f_p/Q_p$ = ________________ Hz.

Draw and solve the circuit of Figure 12.4 for the frequency calculated above.

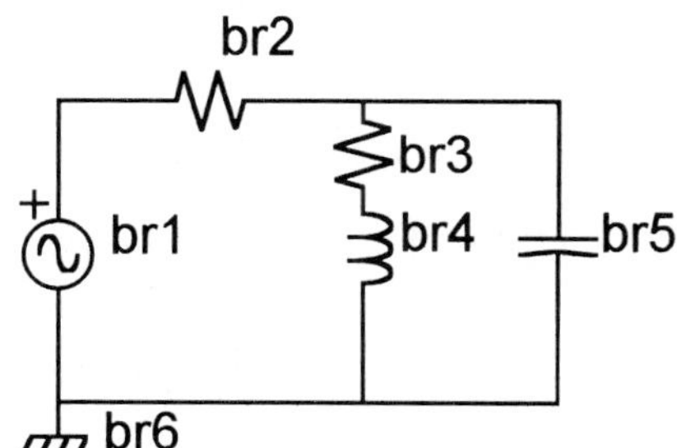

Net List for the Circuit

Br. 1: E = 10 V/0°
Br. 2: R = 10 kΩ
Br. 3: R = 24.95 Ω
Br. 4: L = 7.94 mH
Br. 5: C = 31.8 nF
Br. 6: voltage reference

Figure 12.4

From the solution results determine the following.

The source current = ________________ mA.
The voltage output = V_C = ________________ V.
The inductive reactance of the coil = $X_L = V_L/I_L$ = ________________ Ω.
The Q of the coil = X_L/R_l = ________________ .
The Q_p of the circuit = I_C/I_S = ________________ .
The impedance of the tank circuit is V_C/I_S = ________________ Ω.

Obtain a graph of the voltage output (node 2) over the frequency range of 1 kHz to 100 kHz. From the graph obtain the following.

The resonant frequency is ________________ Hz.
The upper and lower bandwidth frequencies are
f_2 = ________ Hz f_1 = ________ Hz.
The bandwidth is BW = $f_2{-}f_1$ = ________________ Hz.

PROCEDURE Part 2

In the circuit of figure 12.5, the resistor of branch 7 represents the load on the circuit. This loading of the circuit will affect the bandwidth and the voltage output of the circuit.

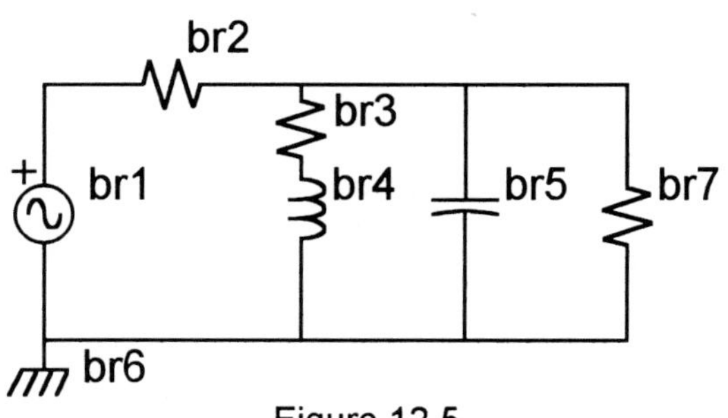

Figure 12.5

Net List for the Circuit

Br. 1: E = 10 V/0°
Br. 2: R = 10 kΩ
Br. 3: R = 24.95 Ω
Br. 4: L = 7.94 mH
Br. 5: C = 31.8 nF
Br. 6: voltage reference
Br. 7: R = 10 kΩ

From the solution results determine the following.

The source current = ________________ mA.
The voltage output = V_C = ________________ V.
The inductive reactance of the coil = $X_L = V_L / I_L$ = ________________ Ω.
The Q of the coil = X_L / Rl = ________________ .
The Q_p of the circuit = I_C / I_S = ________________ .
The impedance of the tank circuit is V_C / I_S = ________________ Ω.

Obtain a graph of the voltage output (node 2) over the frequency range of 1 kHz to 100 kHz. From the graph obtain the following.

The resonant frequency is ________________ Hz.
The upper and lower bandwidth frequencies are
f_2 = ________________ Hz f_1 = ________________ Hz.
The bandwidth is BW = $f_2 - f_1$ = ________________ Hz.

Make the following comparisons for the characteristics of the unloaded and loaded resonant circuits.

The resonant frequency is ________ Hz unloaded and ________ Hz loaded.
The voltage output is ________ V unloaded and ________ V loaded.
The Q_p of the circuit is ________ unloaded and ________ loaded.
The bandwidth is ________ Hz unloaded and ________ Hz loaded.

FURTHER ANALYSIS

What if the coil has a low Q? Will the resonant frequency change? What happens to the bandwidth? Change the resistance of the coil to 100 Ω. Obtain a graph of the frequency response and analyze as in the procedures above.

Frequency - Selective Circuits

OBJECTIVE

To obtain the frequency response of several different types of filters.

THEORY

The subject of filters is very extensive and would require a separate course on its own. Filters are classified as low-pass, high-pass, bandpass, or bandstop and may be active filters incorporating transistors or operational amplifiers or made up of R, L, and C components, in which case they are called passive filters. The active filters incorporate R, L or C elements to obtain the desired filtering effect. Previous exercises have included the R-C and R-L low-pass and high-pass filters and the series and parallel resonant circuits that are used for bandpass and bandstop filters. This exercise will look at other types of passive filter circuits.

PROCEDURE Part 1

The R-C and R-L filters provided low-pass characteristics in which the roll off of the signal was at a rate of −20 dB per decade above the corner frequency (f_c). These are called first order filters. Second order filters provide a roll off of −40 dB per decade. The design requirements for a second order lowpass filter for a given value of resistance are:

$$ C = \frac{\sqrt{2}}{2\pi f_c R} \qquad L = \frac{1}{(2\pi f_c)^2 C} \qquad f_c = \frac{1}{2\pi\sqrt{LC}} $$

For the second order low-pass circuit shown in Figure 13.1, draw the circuit and obtain a Bode plot of the voltage output (node 3) over the frequency range of 1 kHz to 1 MHz.

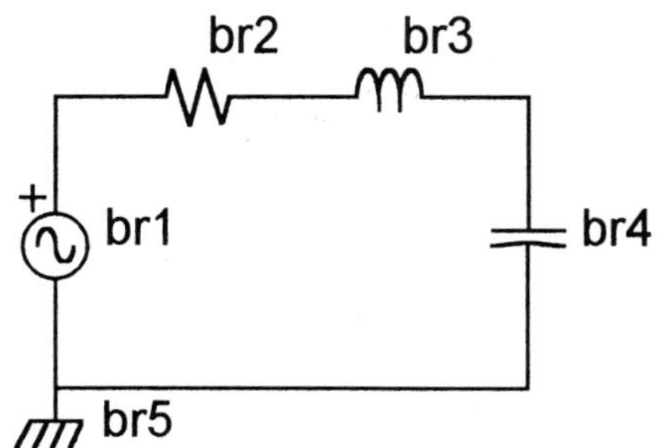

Net List for the Circuit

Br. 1: E = 10 V/0°
Br. 2: R = 1.0 kΩ
Br. 3: L = 11.25 mH
Br. 4: C = 22.5 nF
Br. 5: voltage reference

Figure 13.1

The calculated cutoff frequency is _________________ Hz.

From the graph, the –3 dB frequency is _________________ Hz.

From the graph, the roll off is _________________ dB per decade.

PROCEDURE Part 2

Audio amplifiers are designed to amplify signals in the audio range of frequencies that extend from 20 Hz to 20,000 Hz. Three separate speakers are used to reproduce the sound since any one speaker is incapable of handling the complete range of frequencies without sound degradation. The woofer (bass speaker) is capable of handling the low frequencies and the mid-range speaker and tweeter can reproduce the mid-range and high frequencies.

To separate the music signal into these frequency ranges a crossover network is used consisting of a low-pass, a bandpass and a highpass filter. The circuit of figure 13.2 has been designed for a lowpass corner frequency of 200 Hz, a mid range bandpass of from 200 to 2000 Hz and a highpass corner frequency of 2000 Hz to match the individual speaker frequency response. To provide high-fidelity reproduction of the audio signal, the total power output level of the speaker system must correspond exactly to the power level of the audio signal. For maximum power transfer, the impedance of the network must equal the output impedance of the amplifier at all frequencies.

In the crossover network of Figure 13.2 the source resistance is shown as 8 Ω and each speaker is represented as an 8 Ω resistance. Branch 5 is the woofer, branch 8 is the mid-range speaker and branch 11 is the tweeter. The corresponding filter is associated with each speaker.

The design criteria for the three filters is given below.

for the low-pass $\quad f_L = 200$ Hz $\quad\quad L_L = \dfrac{\sqrt{2}R}{2\pi f_L} \quad\quad C_L = \dfrac{1}{(2\pi f_L)^2 L_L}$

for the mid-range $\quad f_L = 200$ Hz $\quad\quad L_M = \dfrac{R}{2\pi(f_H - f_L)} \quad\quad C_M = \dfrac{1}{(2\pi)^2 f_H f_L L_M}$
and $f_H = 2000$ Hz

for the high-pass $\quad f_T = 2000$ Hz $\quad\quad L_T = \dfrac{\sqrt{2}R}{2\pi f_T} \quad\quad C_T = \dfrac{1}{(2\pi f_T)^2 L_T}$

Draw the complete crossover network shown in Figure 13.2.

Net List for the Circuit

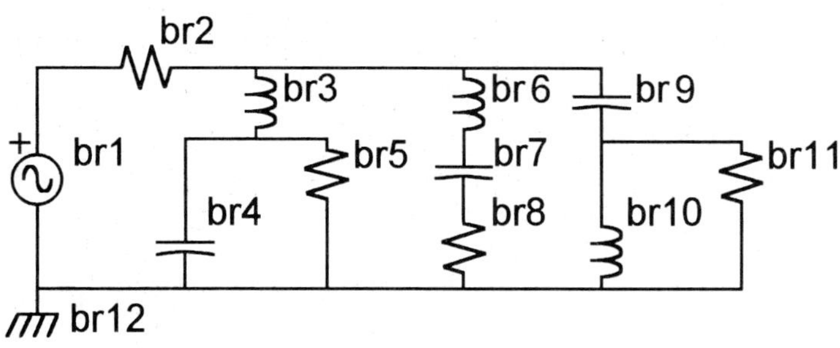

Figure 13.2

Br. 1: E = 20 V/0°
Br. 2: R = 8 Ω
Br. 3: L = 9 mH
Br. 4: C = 70.4 μF
Br. 5: R = 8 Ω
Br. 6: L = 0.707 mH
Br. 7: C = 90 μF
Br. 8: R = 8 Ω
Br. 9: C = 7.04 μF
Br. 10: L = 0.9 mH
Br. 11: R = 8 Ω
Br. 12 voltage reference

Obtain a Bode plot of the voltage across the woofer (node 3) over the frequency range of 10 Hz to 100 kHz and sketch the plot on the graph on Figure 13.3. The program will request the voltage reference value to use for the calculation of the decibels. Enter the value 10 volts.

Obtain a Bode plot of the voltage across the mid-range speaker (node 5) over the same frequency range and sketch the plot on the graph of Figure 13.3.

Repeat the above for the tweeter in branch 11 (voltage at node 6).

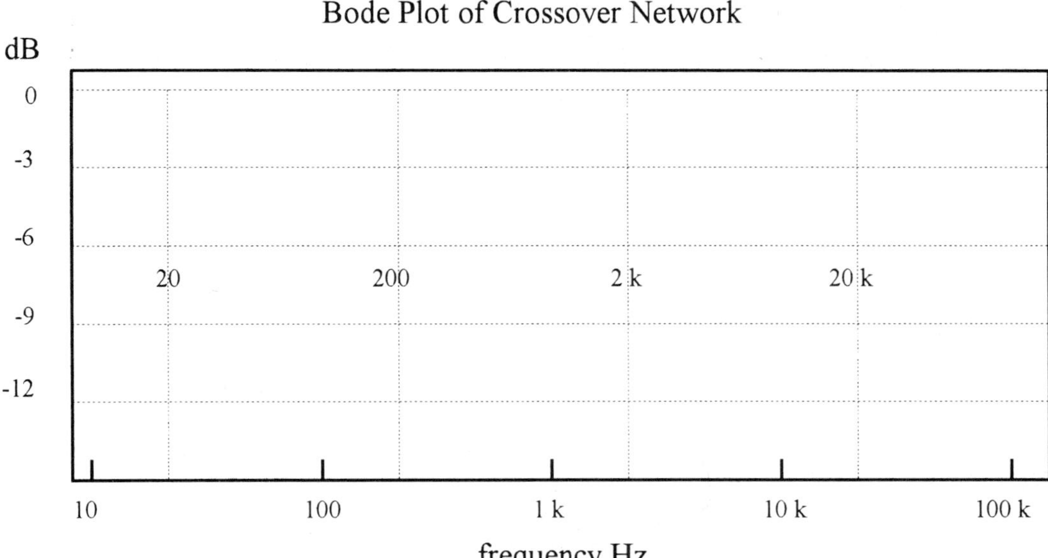

From the Bode plot:

the cutoff frequency for the low-pass circuit is ___________________ Hz

the lower bandwidth frequency for the mid-range speaker is __________ Hz

the upper bandwidth frequency for the mid-range speaker is _________ Hz

the cutoff frequency for the high-pass filter is ________________ Hz

Solve the circuit and calculate the impedance $\mathbf{Z} = V_{node\ 2}/I_{branch\ 2}$ for the following frequencies and calculate the impedance of the network. The impedance should be the same at any frequency.

At 50 Hz the impedance $\mathbf{Z} =$ ___________________ Ω

At 500 Hz the impedance $\mathbf{Z} =$ ___________________ Ω.

At 5 kHz the impedance $\mathbf{Z} =$ ___________________ Ω.

At 10 kHz the impedance $\mathbf{Z} =$ ___________________ Ω.

Obtain a graph of the power delivered from the source (branch 2) over the frequency range of 10 Hz to 100 kHz. Is the power output relatively constant over the complete frequency range? _______________

Comment on the overall performance of the crossover network.

Comment

FURTHER ANALYSIS

There are many types of filters used in electronics. Refer to a text that describes a different type of filter than the types described above, and obtain the frequency response graph of the voltage output over the frequency range of interest. Study the characteristics of the graph, and comment on them.

A Circuit Resonant at All Frequencies

OBJECTIVE

To show that the circuit in Figure 14.1 is resonant at all frequencies.

THEORY

A special case of parallel resonance is shown in the circuit of Figure 14.1 For certain specific values of R, L, and C, the circuit shown can be made resonant at all frequencies. The impedance of the circuit is resistive and of constant value for all frequencies. This circuit finds application in the design of cables, particularly coaxial cables, which are designated by their characteristic impedance Z_o. The value of the characteristic impedance

$$Z_o = \sqrt{\frac{L}{C}} \text{ for values of } R = \sqrt{\frac{L}{C}}$$

PROCEDURE

Draw the circuit using *Breadboard* for the values given in the net list. Solve the circuit for the frequencies shown in Table 14.1. From the solution results, complete the table and calculate the characteristic impedance for each frequency.

Net List for the Circuit

Br. 1: E = 10 V/0°
Br. 2: R = 50 Ω
Br. 3: C = 1.0 nF
Br. 4: R = 50 Ω
Br. 5: L = 2.5 µH
Br. 6: voltage reference

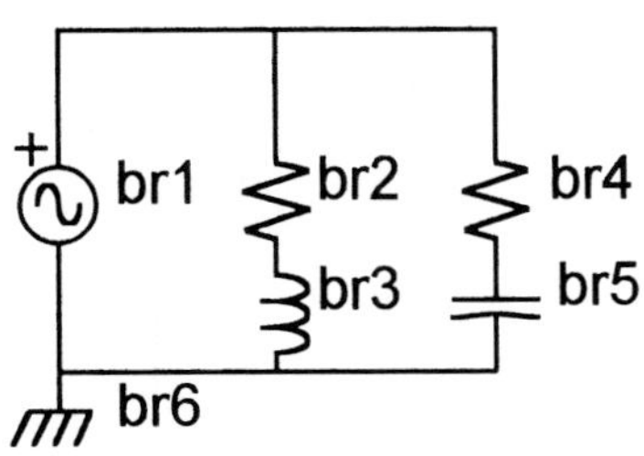

Figure 14.1

Table 14.1

Frequency	Source current	Phase angle	$Z_o = E/I$
1.0 kHz			
10 kHz			
100 kHz			
1.0 MHz			

The characteristic impedance of the circuit is

$$Z_o = E/I_T = \underline{\hspace{3cm}} \angle \underline{\hspace{3cm}} ^\circ \ \Omega.$$

The calculated characteristic impedance is $\underline{\hspace{3cm}}$ Ω.

FURTHER ANALYSIS

Obtain a graph of the current from the source over a frequency range of 1 kHz to 1 MHz. Is the current magnitude constant and the phase angle 0°?

A Constant Source Current Under Varying Load

OBJECTIVE

To observe that a constant magnitude source current is achieved for a varying resistive load in the circuit of Figure 15.1 and to introduce polar (locus) plots.

THEORY

By choosing values of capacitance and inductance such that X_C equals $2X_L$ in the circuit of Figure 15.1, the source current will be constant in magnitude regardless of the value of the load resistor. However, the phase angle of the source current will vary from +90° to -90°. To graphically represent this result, a polar plot of the current phasor at different values of load resistance is to be drawn.

PROCEDURE

To prove that the source current is constant in magnitude, the circuit of Figure 15.1 is solved using *Breadboard* for the range of load resistor values shown in Table 15.1 at a frequency of 60 Hz. Complete the table for the magnitude and phase angle of the source current ($I_{\text{branch 1}}$) and the load current ($I_{\text{branch 4}}$).

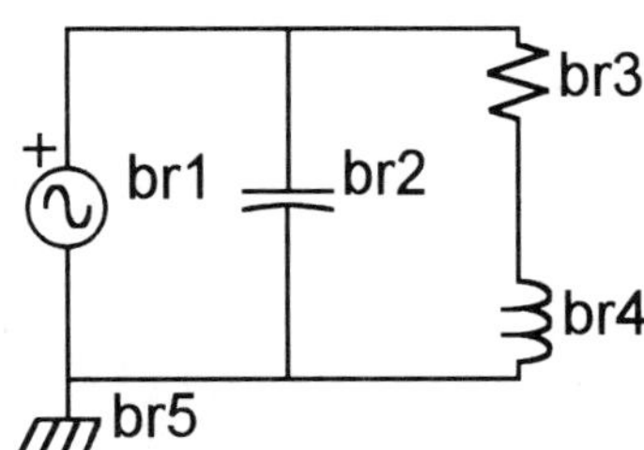

Figure 15.1

Net List for the Circuit

Br. 1: E = 230/0° V
Br. 2: C = 602.8 μF
Br. 3: R = 1.0 Ω
Br. 4: L = 5.836 mH
Br. 5: voltage reference

Table 15.1

$R_{\text{branch 4}}$	$I_{\text{branch 1}}$	Phase angle	$I_{\text{branch 4}}$
0.1 Ω			
1 Ω			
10 Ω			
100 Ω			
1 k Ω			
10 k Ω			

A locus diagram is a plot of a phasor of current or voltage in a circuit as its value changes due to the variation of a parameter of the circuit. The end points of the phasors trace a path (locus) of the phasor. The advantage of a phasor plot is that the magnitude and phase angle variation of the phasor can be seen together in one graph.

Draw a polar graph (locus diagram) approximately to scale for the values of source current obtained. The example current phasor shown in Figure 15.2 was obtained for a load resistance of 5 Ω .

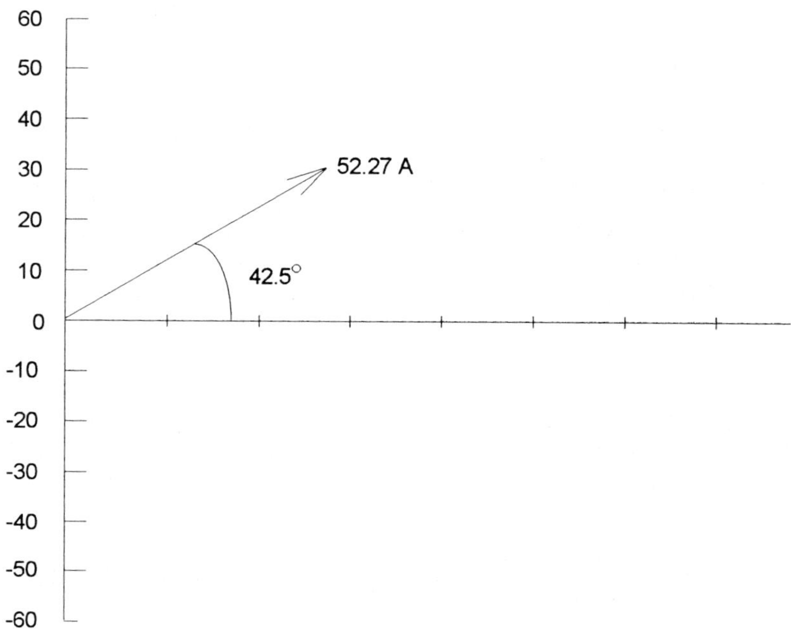

Figure 15.2   Locus Diagram of Source Current

Discuss how this circuit could be used in the operation of an electric-arc furnace for melting scrap steel where the resistance of the arc may vary from a short circuit to an open circuit.

Discussion

FURTHER ANALYSIS

Sketch a locus diagram of the current in the series resonant circuit in Procedure Part 1 of AC Exercise 11.

Phase Shift Trigger Circuit

16

AC Exercise

OBJECTIVE

To obtain the voltage output of the phase shift circuit shown in Figure 16.1.

THEORY

A variable phase shift trigger circuit is employed to trigger silicon-controlled rectifiers (SCRs) in power-rectification applications. Ideally, the trigger signal should be constant in magnitude and variable over a phase angle range of $0°$ to $180°$. The trigger circuit in Figure 16.1 performs this function well with the trigger voltage taken across the resistor shown as branch 6 in the figure. Branch 4 represents the variable resistor (a potentiometer), which determines the phase shift.

PROCEDURE

Draw the circuit shown in Figure 16.1 for the component values given in the net list. Solve for the voltage across branch 6 for a frequency of 60 Hz. Complete the table for the remaining values of branch 4 resistance.

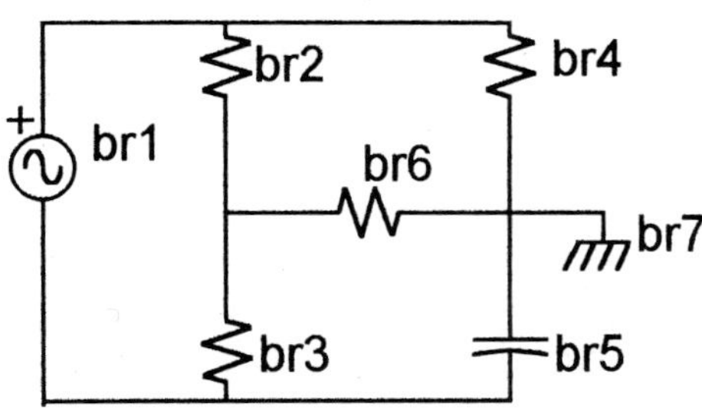

Figure 16.1

Net List for the Circuit

Br. 1: $E = 120/0°$ V
Br. 2: $R = 240\ \Omega$
Br. 3: $R = 240\ \Omega$
Br. 4: $R = 500\ \Omega$
Br. 5: $C = 5.0\ \mu F$
Br. 6 : $R = 10\ k\Omega$
Br. 7: voltage reference

Table 16.1

$R_{\text{branch 4}}$	$V_{\text{branch 6}}$	Phase angle
10 kΩ		
1.0 kΩ		
500 Ω		
100 Ω		
10 Ω		
1 Ω		

Using the results in the table for the trigger voltage and its phase angle, draw a phasor plot (locus diagram) for the voltage across branch 6.

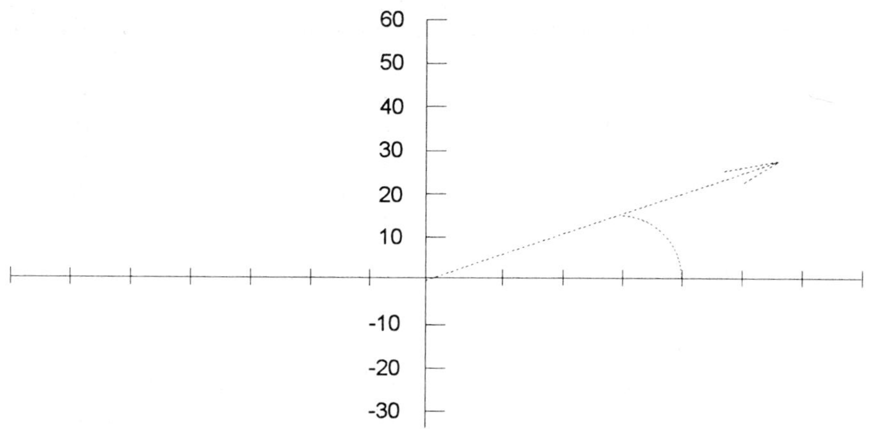

Figure 16.2 Locus Diagram of Trigger Voltage

From the results of the polar plot comment on the variation of the trigger voltage.

Comment

FURTHER ANALYSIS

A simple trigger circuit may consist of just an *R-C* series combination (Figure 16.3). Model the circuit and test for the voltage magnitude and phase shift of the output voltage ($V_{\text{branch 4}}$) using the same procedure as before.

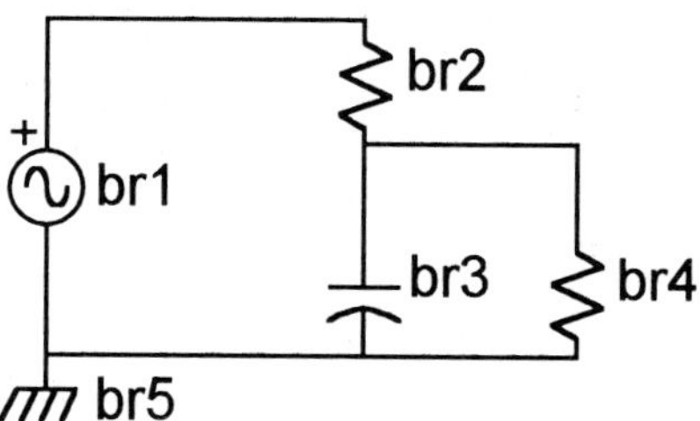

Figure 16.3

Net List for the Circuit

Br. 1: E = 10 V/0°
Br. 2: R = 1 kΩ
Br. 3: C = 0.1 μF
Br. 4: R = 100 kΩ
Br. 5: voltage reference

Three-Phase Three-Wire WYE-Connected System

OBJECTIVE

To analyze a three-phase three-wire WYE-connected circuit with a balanced and an unbalanced load.

THEORY

Three-phase systems consist of three equal AC voltages, electrically displaced by 120°. The transmission of three-phase power offers a net reduction in the size of conductors required when compared to other systems. In addition, the physical size and cost of three-phase transformers and motors are reduced. Three-phase motors have better torque characteristics than single-phase due to the rotating magnetic field produced by the three-phase currents in the motor. A further advantage is that the instantaneous power of a balanced three-phase system is constant.

The three-phase WYE-connected system shown in Figure 17.1 provides a two-voltage system, since the magnitude of the line-to-line voltage equals $\sqrt{3}$ times the line-to-neutral voltage. The phase current in a WYE system is the same as the line current.

$$I_{line} = I_{phase}$$

$$V_{line} = \sqrt{3}\, V_{phase}$$

The total power delivered to the load is the sum of the power in each of the phases.

$$P_T = P_{AN} + P_{BN} + P_{CN}$$

If the load is **balanced**, the total power can be calculated as $P_T = 3P_{phase} = 3V_P I_P \cos\theta$ or as $P_T = \sqrt{3} V_L I_L \cos\theta$ where θ is the load impedance angle.

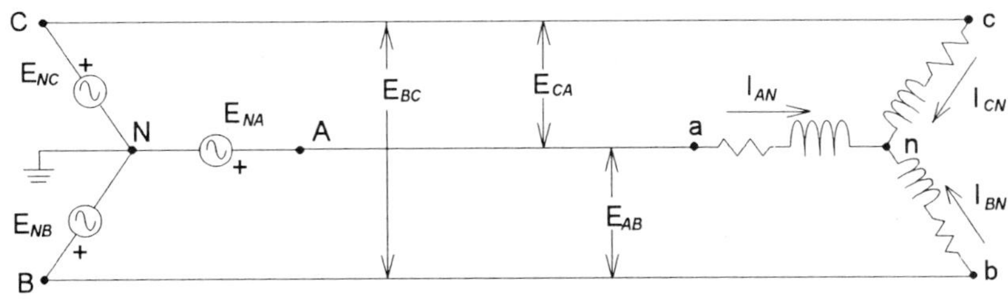

Figure 17.1

PROCEDURE Part 1

The circuit of Figure 17.2 represents a three-phase, three-wire, WYE-connected alternator providing power to an **balanced** WYE-connected load. A three-phase three-wire network must have a balanced load as there is no neutral to carry the unbalance current. If the load is unbalanced an undesirable condition known as a "floating neutral" will result.

An analysis of the current in the lines and the total power in the load is to be obtained. Use *Breadboard* to solve for the currents in the circuit of Figure 17.2.

Mark on the circuit diagram the current in each phase of the source and the load.

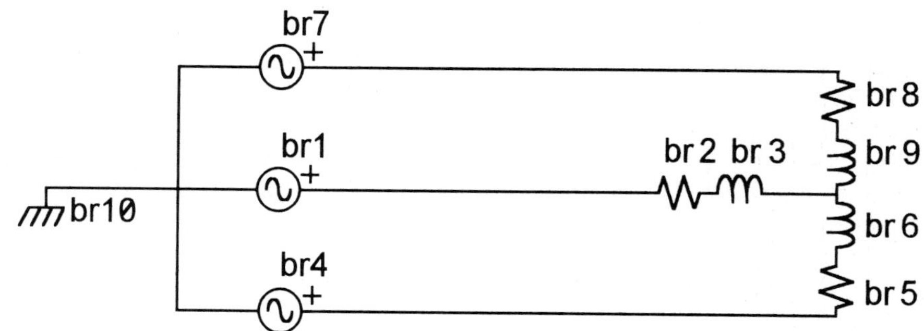

Figure 17.2

Net List for the Circuit

Br. 1: $E = 120$ V/0°   Br. 6: $L = 21.22$ mH
Br. 2: $R = 6\ \Omega$ Br. 7: $E = 120$ V/120°
Br. 3: $L = 21.22$ mH   Br. 8: $R = 6\ \Omega$
Br. 4: $E = 120$ V/-120°   Br. 9: $L = 21.22$ mH
Br. 5: $R = 6\ \Omega$ Br. 10: voltage reference

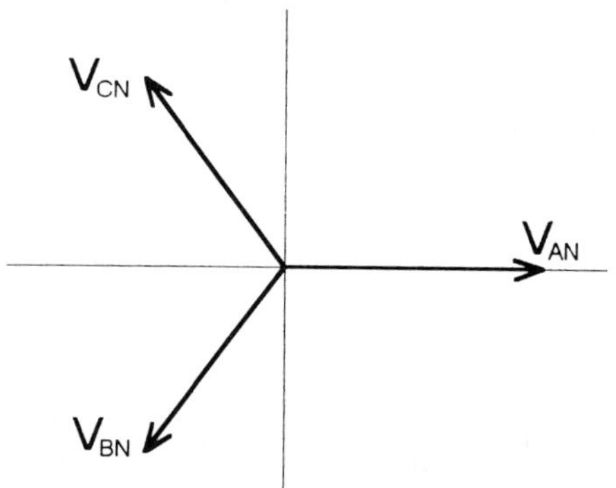

Phasor Diagram for
Balanced Three-Wire WYE-Load

On the adjacent phasor diagram draw the phase currents approximately to scale. Show graphically that the phasor sum of the phase currents is zero.

Calculate the sum of the three phase currents to prove that their sum is zero.

$$I_{AN} + I_{BN} + I_{CN} = \underline{\hspace{2cm}} \text{ A}$$

From the solution results of the power in each phase of load, find the total power as the sum of the phase powers.

$$P_T = P_A + P_B + P_C = \rule{4cm}{0.4pt} \quad W$$

From the solution results the total power is \rule{3cm}{0.4pt} W

Calculate the total power as $P_T = V_L I_L \cos\theta = $ \rule{3cm}{0.4pt} W

Calculate the total power as $3V_P I_P \cos\theta = $ \rule{3cm}{0.4pt} W

PROCEDURE Part 2

Edit the circuit and add 1 MΩ resistors (voltmeters) across each line as branches 11, 12, and 13 as in Figure 17.3 to measure the line voltages V_{AB}, V_{BC}, and V_{CA}.

The magnitude of the line voltages V_{AB}, V_{BC}, and V_{CA} is \rule{3cm}{0.4pt} V.

By calculation the line voltages are $V_L = \sqrt{3}V_P = $ \rule{3cm}{0.4pt} V.

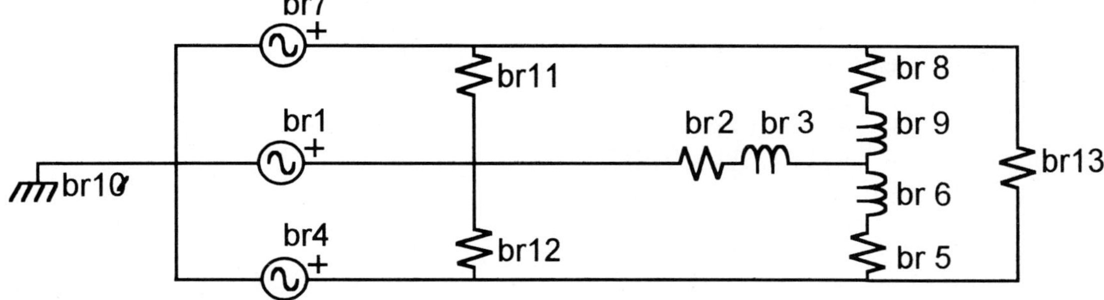

Figure 17.3

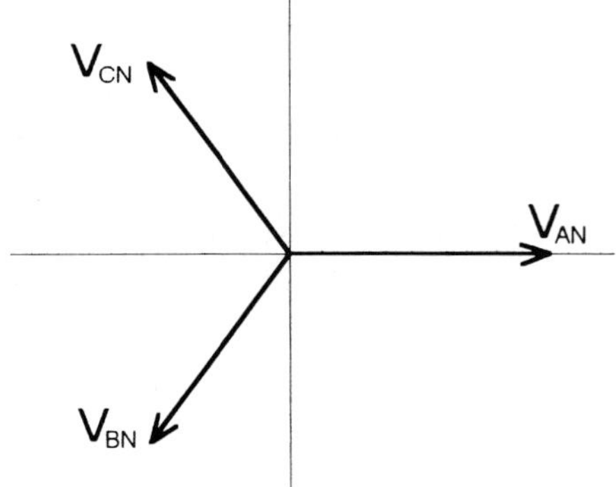

Phasor Diagram for
Three-Phase WYE-WYE System

On the adjacent phasor diagram draw the line voltages approximately to scale to show the 30° shift from the phase voltages.

Show graphically the following relationships.

$$V_{AB} = V_{AN} - V_{BN}$$
$$V_{BC} = V_{BN} - V_{CN}$$
$$V_{CA} = V_{CN} - V_{AN}$$

FURTHER ANALYSIS

If the neutral wire of an **unbalanced** three-wire WYE-connected load becomes disconnected, an undesirable condition known as a "floating neutral" results. This condition causes the voltages across each phase of the load to become unbalanced. Draw the circuit of Figure 17.4 and solve for the voltages across each of the load impedances. Branches 10, 11, and 12 represent the voltmeters across the load impedances. Mark on the circuit diagram the current in each phase of the load.

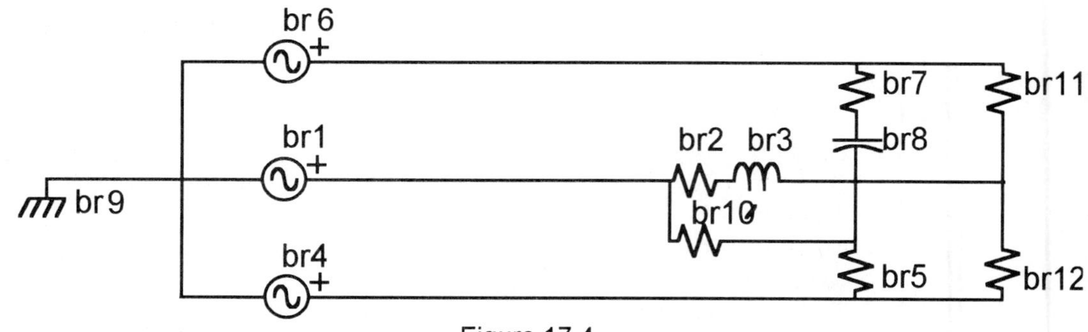

Figure 17.4

Net List for the Circuit

Br. 1: $E = 120$ V/0°
Br. 2: $R = 26$ Ω
Br. 3: $L = 39.8$ mH
Br. 4: $E = 120$ V/-120°
Br. 5: $R = 40$ Ω
Br. 6: $E = 120$ V/120°

Br. 7: $R = 15$ Ω
Br. 8: $C = 177$ μF
Br. 9: voltage reference
Br. 10: $R = 1$MΩ
Br. 11: $R = 1$MΩ
Br. 12: $R = 1$ MΩ

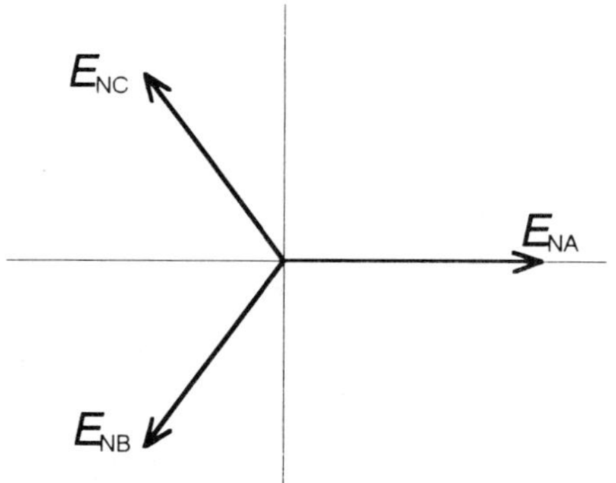

Phasor Diagram for Unbalanced
Three-Wire WYE Load

Draw the phasor diagram for the voltage across each branch of the load.

Calculate the phasor sum of the phase voltages.

$$V_{AN} + V_{BC} + V_{CN} = \text{\_\_\_\_\_\_\_\_} \text{ V}$$

Calculate the phasor sum of the load currents. Is the sum zero?

$$I_{AN} + I_{BN} + I_{CN} + \text{\_\_\_\_\_\_\_\_} \text{ A}$$

Three-Phase Four-Wire WYE-Connected System

OBJECTIVE

To study the operation of a four wire WYE–connected distribution system.

THEORY

Three-wire three-phase systems are used for the transmission of electrical power to take advantage of the efficiencies of the system. For the distribution of electrical power in utilities and industrial plants, the three phase system with a neutral is used to obtain a two voltage system. The line-to-neutral voltage is $1/\sqrt{3}$ of the line-to-line voltages. The 208/120 V system, for example, provides 208 volts line-to-line for motor loads and 120 volts line-to-neutral for lighting loads. The four-wire system is shown in Figure 18.1. The phase loads are kept equal as must as possible but if the loads are unequal the unbalance current is returned to the source through the neutral wire.

The following relationships apply for the four wire system .

$$V_{line} = \sqrt{3}\,V_{phase}$$

$$I_{line} = I_{phase}$$

$$\boldsymbol{I}_{AN} + \boldsymbol{I}_{BN} + \boldsymbol{I}_{CN} = \boldsymbol{I}_{N}$$

$$P_{T} = P_{AN} + P_{BN} + P_{CN}$$

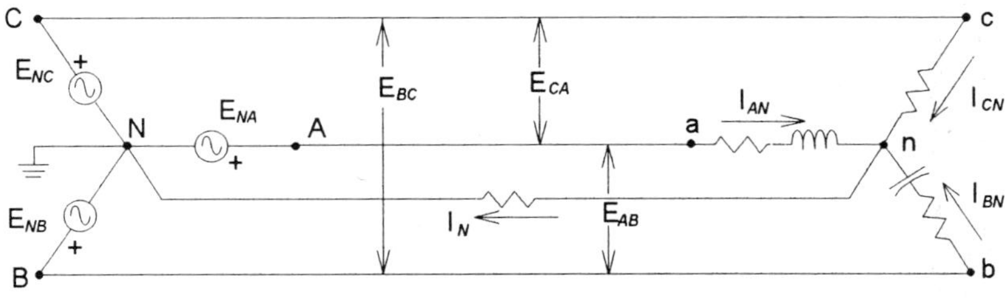

Figure 18.1

The circuit of Figure 18.2 represents a power source (an alternator or transformer) providing power to an **unbalanced** three-phase load. The resistor in the neutral line (branch 9) represents an ammeter to measure the current in the line.

PROCEDURE Part 1

Solve the circuit shown in Figure 18.2 using *Breadboard* for a frequency of 60 Hz.

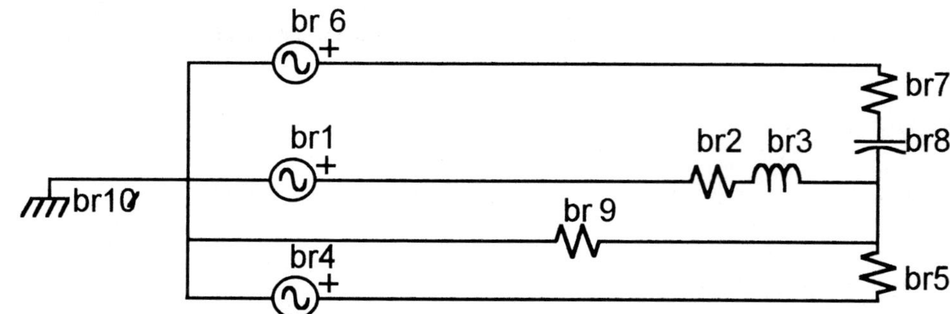

Figure 18.2

Net List for the Circuit

Br. 1: E = 120 V/0°   Br. 6: E = 120 V/120°
Br. 2: R = 26 Ω   Br. 7: R = 42.24 Ω
Br. 3: L = 39.8 mH   Br. 8: C = 62.53 µF
Br. 4: E = 120 V/-120°   Br. 9: R = 0.1 mΩ
Br. 5: R = 40 Ω   Br. 10: voltage reference

Record the current of each phase and the neutral on the circuit diagram.

Calculate the neutral current as the phasor sum of the phase currents.

$$I_N = I_{AN} + I_{BN} + I_{CN} = \underline{\qquad} \angle \underline{\qquad} °\ A$$

From the solution the neutral current is ________ $\angle$ _______ ° A.

The total power in the load is the sum of the power in each phase.

$$P_T = P_{AN} + P_{BN} + P_{CN} \underline{\hspace{5cm}} W.$$

From the solution the total power absorbed in the resistors is __________ W.

From the solution the power delivered by the sources verify that

$$P_T = P_{NA} + P_{NB} + P_{NC} = \underline{\hspace{5cm}} W.$$

The total power delivered from the source may be calculated as

$$P_T = E_{NA} I_{NA} \cos\theta + E_{NB} I_{NB} \cos\phi + E_{NC} I_{NC} \cos\varphi = \underline{\hspace{2.5cm}} W.$$

Is the value of total power the same in each case above? ________

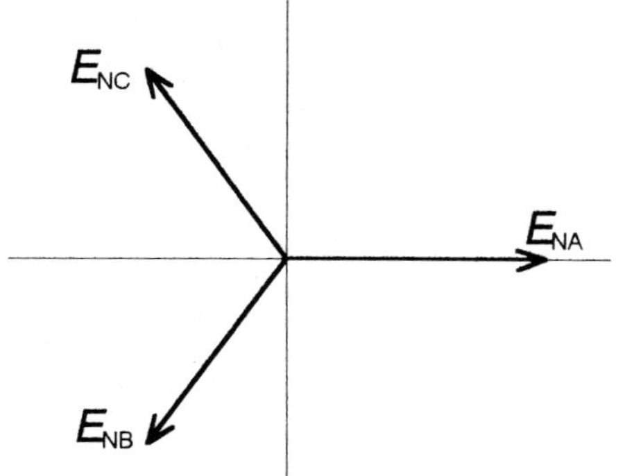

On the adjacent phasor diagram draw the current of each phase of the load and the neutral current.

Phasor Diagram for
Unbalanced Four-Wire WYE Load

PROCEDURE Part 2

The phase sequence for the three-phase system in Figure 18.2 is an ABC sequence since the phase angles of the source voltages are 0°, -120°, and -240° for E_{AN}, E_{BN}, and E_{CN}. If the sequence is changed to CBA the currents in the three phases will have different values.

Edit the circuit of Figure 18.2 and change the phase sequence to CBA as shown in Figure 18.3

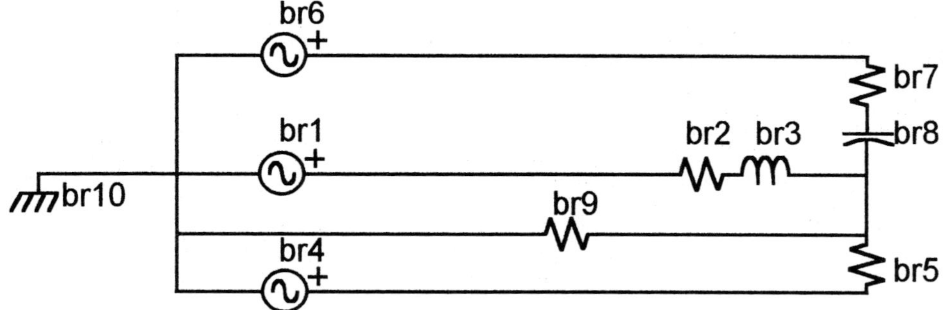

Figure 18.3

Net List for the Circuit

Br. 1: E = 120 V/0°   Br. 6: E = 120 V/-120°
Br. 2: R = 26 Ω   Br. 7: R = 42.24 Ω
Br. 3: L = 39.8 mH   Br. 8: C = 62.53 μF
Br. 4: E = 120 V/120°   Br. 9: R = 0.1 mΩ
Br. 5: R = 40 Ω   Br. 10: voltage reference

Mark on the circuit diagram of Figure 19.3 the current in each branch and the neutral current.

Calculate the neutral current as the phasor sum of the phase currents.

$$I_N = I_{AN} + I_{BN} + I_{CN} = \underline{\hspace{2cm}} \angle \underline{\hspace{1.5cm}} °\ A$$

From the solution the neutral current is $\underline{\hspace{2cm}} \angle \underline{\hspace{1.5cm}} °\ A.$

The total power is the sum of the power in each phase.

$$P_T = P_{AN} + P_{BN} + P_{CN} \underline{\hspace{6cm}} W.$$

From the solution the total power absorbed in the resistors is $\underline{\hspace{3cm}}$ W.

From the solution for power in the source verify that

$$P_T = P_{NA} + P_{NB} + P_{NC} = \underline{\hspace{5cm}} W.$$

The total power delivered from the source may be calculated as

$$P_T = E_{NA} I_{NA} \cos\theta + E_{NB} I_{NB} \cos\phi + E_{NC} I_{NC} \cos\varphi = \underline{\hspace{2cm}} W.$$

Is the value of total power the same in each case above? $\underline{\hspace{2cm}}$

Write a summary in point form of the characteristics of the three-phase four-wire system.

Summary

Three-Phase WYE-DELTA Connected System

OBJECTIVE

To analyze a three-phase system with the source voltages connected WYE and the load impedances connected DELTA.

THEORY

Three-phase loads may be connected in a DELTA configuration as shown in Figure 19.1. The line and phase voltages are equal since the impedances of the load are connected from line-to-line. Although no neutral wire is possible with such a system, the loads may be balanced or unbalanced. The **line currents** produced will sum to zero, obviating the need for a neutral connection. Note that the sum of the phase currents will **not** sum to zero if the load is unbalanced.

$$V_{line} = V_{phase}$$

$$I_A = I_{AB} - I_{CA}$$
$$I_B = I_{BC} - I_{AB}$$
$$I_C = I_{CA} - I_{BC}$$

$$P_T = P_{AB} + P_{BC} + P_{CA}$$

For a balanced DELTA load

$$I_{line} = \sqrt{3} I_{phase} \text{ in magnitude}$$

$$P_T = 3P_{phase} = \sqrt{3} V_L I_L \cos\theta \qquad \text{where } \theta \text{ is the load impedance angle}$$

For a balanced load the line current is shifted by 30° as is shown in the following procedure part 1.

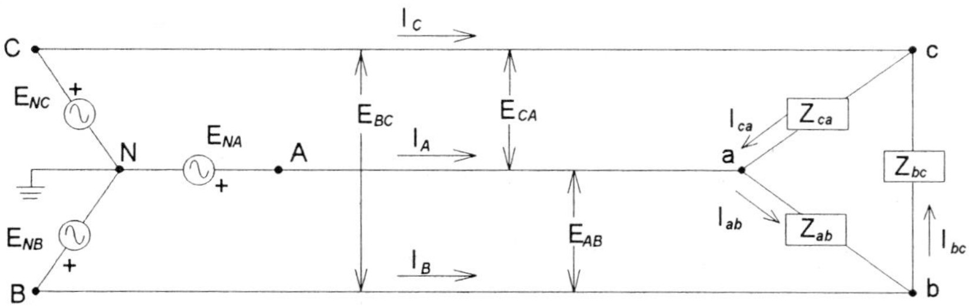

Figure 19.1

PROCEDURE Part 1

Resistors have been chosen as the load for the balanced WYE-DELTA connected system shown in Figure 19.2 in order to observe a 30° phase shift of the line currents when referenced to the phase voltages of the source. Draw and solve the circuit for 60 Hz.

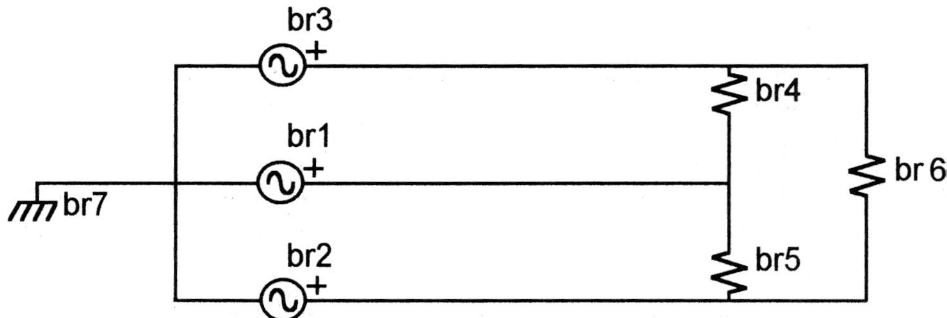

Figure 19.2

Net List for the Circuit

Br. 1: E = 120 V/0°   Br. 5: R = 20.8 Ω
Br. 2: E = 120 V/-120°   Br. 6: R = 20.8 Ω
Br. 3: E = 120 V/120°   Br. 7: voltage reference
Br. 4: R = 20.8 Ω

Mark the phase currents and the line currents on the circuit diagram.

Does the magnitude of the line current equal $\sqrt{3}$ times the phase current?

$$I_{line} = \sqrt{3}I_{phase} = \underline{\hspace{3cm}} \text{ A}$$

Calculate $I_A = I_{ab} - I_{bc} = \underline{\hspace{2cm}} \angle \underline{\hspace{1cm}}°$ A. Solution result for I_A is $\underline{\hspace{2cm}} \angle \underline{\hspace{1cm}}°$A.

Calculate the phasor sum of the line currents. $I_A + I_B + I_C = \underline{\hspace{2cm}} \angle \underline{\hspace{1cm}}°$A.

On the adjacent phasor diagram draw the phase currents and the line currents.

Graphically show that $I_A = I_{ab} - I_{bc}$.

Graphically show the line voltage $E_{AB.}$

Is there a 30° shift in the line current? $\underline{\hspace{2cm}}$

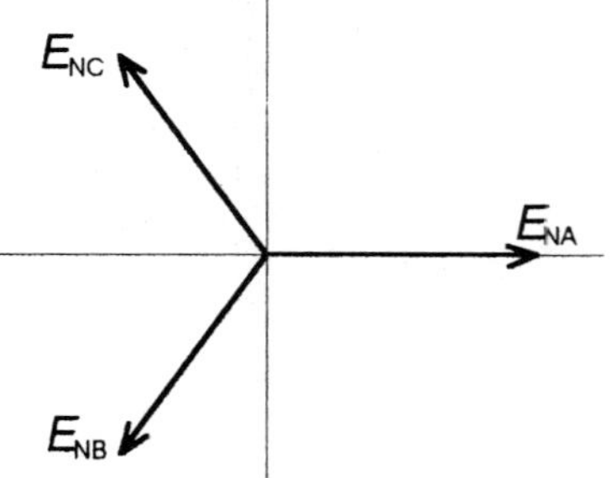

Phasor Diagram for
Balanced DELTA load

From the solution results, the sum of the power in each of the load resistors is

$$P_T = P_{AB} + P_{BC} + P_{CA} = \text{\_\_\_\_\_\_\_\_\_\_\_\_\_\_\_\_\_\_} \text{ W.}$$

From the solution results, sum the power delivered by each phase of the source.

$$P_{NA} + P_{NB} + P_{NC} = \text{\_\_\_\_\_\_\_\_\_\_\_\_\_\_\_\_\_\_} \text{ W}$$

PROCEDURE Part 2

The circuit of Figure 19.3 represents a WYE connected alternator or transformer supplying an unbalanced DELTA connected load.

Draw the circuit using *Breadboard* and solve the circuit for 60 Hz. Mark on the circuit diagram the current in each phase of the load, and the line currents.

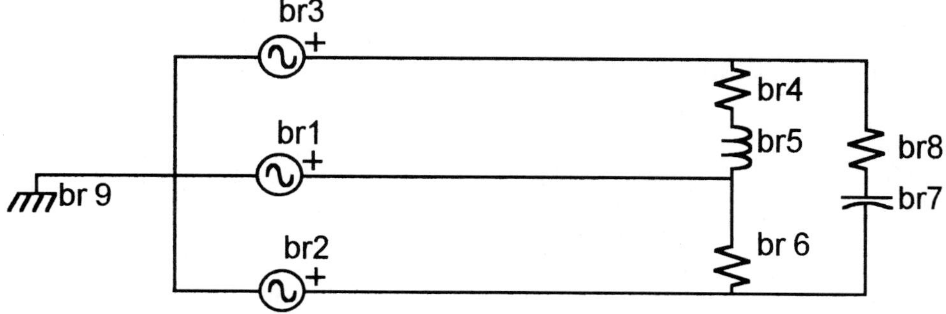

Figure 19.3

Net List for the Circuit

Br. 1: E = 120 V/0°   Br. 6: R = 10.39 Ω
Br. 2: E = 120 V/-120°   Br. 7: C = 120.6 μF
Br. 3: E = 120 V/120°   Br. 8: R = 22 Ω
Br. 4: R = 18 Ω   Br. 9: voltage reference
Br. 5: L = 27.57 mH

Draw the phase and line currents on the adjacent phasor diagram approximately to scale.

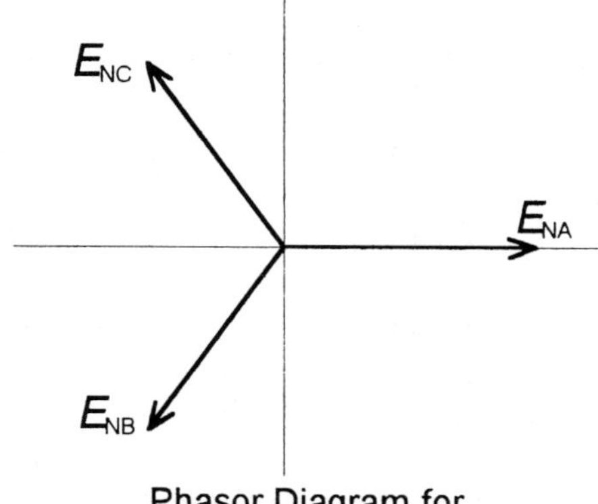

Phasor Diagram for
Unbalanced Delta Load

Calculate the line currents by phasor addition.

$$I_A = I_{AB} - I_{CA} = \underline{\hspace{3cm}} \angle \underline{\hspace{2cm}}^\circ \text{ A}$$
$$I_B = I_{BC} - I_{AB} = \underline{\hspace{3cm}} \angle \underline{\hspace{2cm}}^\circ \text{ A}$$
$$I_C = I_{CA} - I_{BC} = \underline{\hspace{3cm}} \angle \underline{\hspace{2cm}}^\circ \text{ A}$$

Calculate the phasor sum of the line currents. Is the sum zero amperes? \underline{\hspace{2cm}}

$$I_A + I_B + I_C = \underline{\hspace{4cm}} \angle \underline{\hspace{2cm}}^\circ \text{ A}$$

The results of an analysis may be checked by determining if the power delivered by source equals the power dissipated in the load. Perform a power balance check on the results of the analysis for the unbalanced DELTA system below.

From the solution results the sum of the power in each of the load resistors is

$$P_T = P_{AB} + P_{BC} + P_{CA} = \underline{\hspace{5cm}} \text{ W.}$$

From the solution results, sum the power delivered by each phase of the source.

Power in phase NA = \underline{\hspace{5cm}} W
Power in phase NB = \underline{\hspace{5cm}} W
Power in phase NC = \underline{\hspace{5cm}} W
Total power from the source P_T = \underline{\hspace{5cm}} W

From the *Breadboard* solution, the total power dissipated in the resistors is

$$P_T = \underline{\hspace{5cm}} \text{ W.}$$

FURTHER ANALYSIS

Solve a problem in your text for a WYE-DELTA system and then use *Breadboard* to confirm your results.

Three-Phase DELTA-DELTA Connected System

OBJECTIVE

To analyze a three-phase DELTA-DELTA connected circuit for the line and phase currents.

THEORY

A three-phase source may be connected in a DELTA configuration. This configuration appears to be a short circuit connection but recall that the sum of the three phase voltages around the loop is zero and no circulating current will flow. The same advantages accrue for a DELTA connected power distribution system as for the WYE connected system. In the DELTA system however, no neutral point is available.

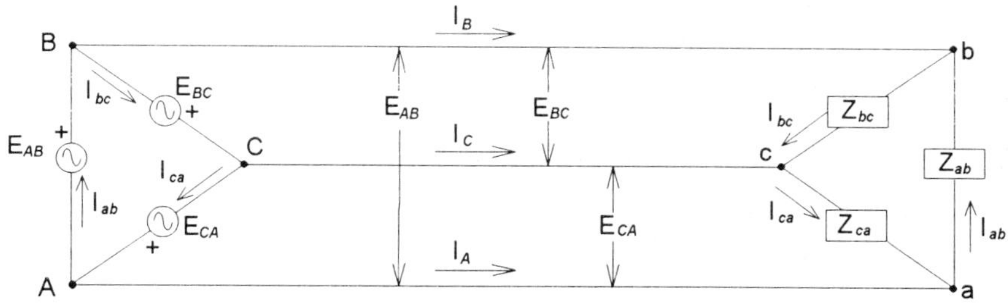

Figure 20.1

The relationships below apply to the DELTA system.

$$V_{line} = V_{phase}$$

$$I_A + I_B + I_C = 0$$

$$P_T = P_{AB} + P_{BC} + P_{CA}$$

For a balanced load

$$I_{line} = \sqrt{3}I_{phase} \qquad \text{in magnitude with a } 30° \text{ shift}$$

$$P_T = 3P_{phase} = \sqrt{3}V_L I_L \cos\theta \qquad \text{where } \theta \text{ is the load impedance angle}$$

As there is no neutral line, the phasor sum of the **line** currents must add to zero whether the load is balanced or unbalanced.

PROCEDURE

The three-phase circuit of Figure 20.2 represents a DELTA-connected source and an **balanced** DELTA–connected load. The system does not have a physical neutral point, but source voltages with their phase angles are specified with reference to a virtual neutral. The program requires that an artificial neutral be created to which it may reference the phase voltages specified. This is done simply by including the three resistors shown within the DELTA of the sources as branches 4, 5, 6, and 7 in Figure 20.2. The value of these resistors should be large in order not to affect the solution values for the circuit. The solution results of the voltages across these resistors are the voltages of an equivalent WYE-connected source.

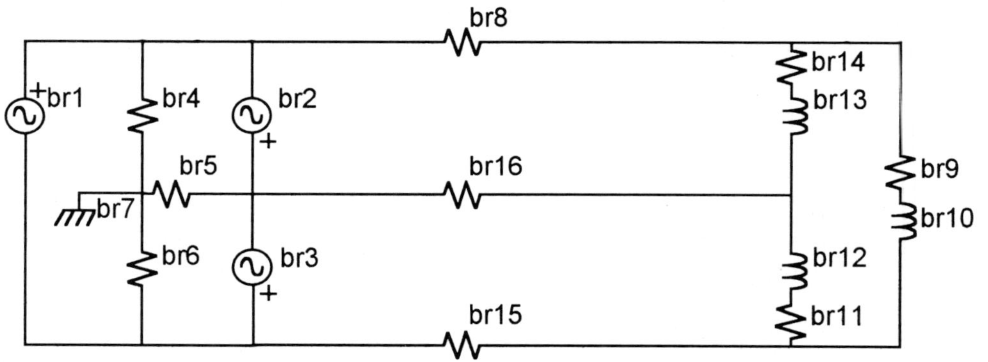

Figure 20.2

In order to have the solution results include the line currents, low-valued resistors representing ammeters or the resistances of the distribution lines, are placed in each of the lines so that the program will solve for the current in these branches.

Net List for the Circuit

Br. 1: E = 120 V/0°	Br. 9: R = 21.2 Ω
Br. 2: E = 120 V/-120°	Br. 10: L = 56 mH
Br. 3: E = 120 V/120°	Br. 11: R = 21.2 Ω
Br. 4: R = 10 kΩ	Br. 12: L = 56 mH
Br. 5: R = 10 kΩ	Br. 13: L = 56 mH
Br. 6: R = 10 kΩ	Br. 14: R = 21.2 Ω
Br. 7: voltage reference	Br. 15: R = 0.01 Ω
Br. 8: R = 0.01 Ω	Br. 16: R = 0.01 Ω

Draw and solve the circuit for the DELTA-connected three-phase circuit of Figure 20.2 for a frequency of 60 Hz. Note the polarities of the voltage sources.

Mark on the circuit diagram the current in each phase and the line currents.

From the solution results, calculate the sum of the line currents by phasor addition.

$$I_A + I_B + I_C = \underline{\hspace{3cm}} \angle \underline{\hspace{2cm}} °\ A$$

The magnitude of the line currents are

$$I_{line} = \sqrt{3}I_{phase} = \underline{\hspace{4cm}} A.$$

The current in each phase of the load is $\underline{\hspace{3cm}} \angle \underline{\hspace{1.5cm}} A.$

The current in each phase of the source is $\underline{\hspace{3cm}} \angle \underline{\hspace{1.5cm}} A$

Comment on the currents from each source compared to the currents in each phase of the load.

Comment

Using the solution results draw the phasor diagram for the phase currents, and the line currents approximately to scale.

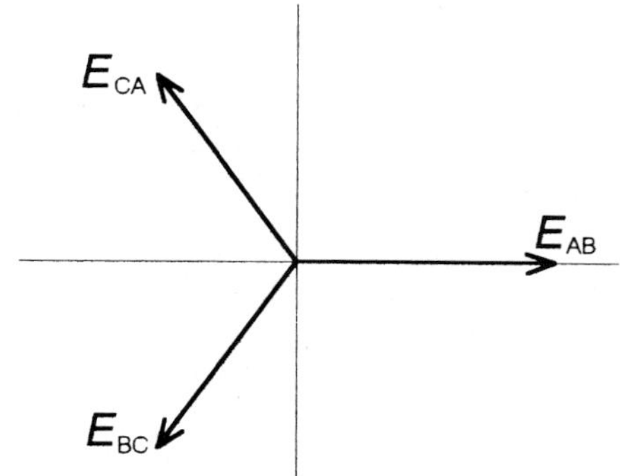

Phasor Diagram for a
Balanced DELTA-DELTA System

Show graphically that the phasor sum of any two phase currents equals the current in the line that is connected to the two phases.

Calculate the line current using phasor algebra.

$$I_A = I_{AB} - I_{BC} = \underline{\hspace{3cm}} A$$

The power in each phase of the load may be calculated as

$$P_{phase} = V_P I_P \cos\theta = \underline{\hspace{4cm}} W.$$

The *Breadboard* result for power in each resistor of the load is $\underline{\hspace{3cm}} W.$

The total power in the load can be calculated as

$$P_T = \sqrt{3} V_P I_P \cos\theta = \underline{\hspace{5cm}} W.$$

The *Breadboard* result for total power in the resistors is $\underline{\hspace{3cm}} W.$

PROCEDURE Part 2

The DELTA load may also be an **unbalanced** load. The phasor sum of the **line** currents will add to zero. To solve for the currents in the source would require the method of mesh analysis, as they are not the same as the currents in each phase of the load. Breadboard facilitates the solution for the source currents.

Draw and solve the circuit shown in Figure 20.3 for a frequency of 60 Hz.

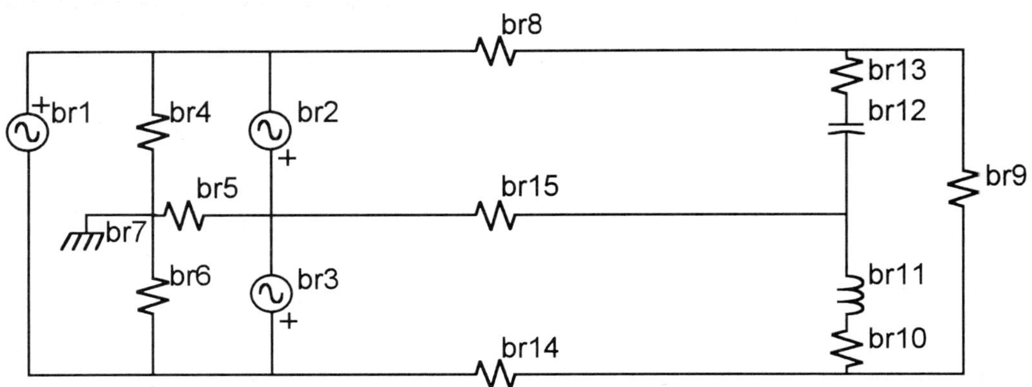

Figure 20.3

Net List for the Circuit

Br. 1: E = 120 V/0°   Br. 8: R = 0.01 Ω
Br. 2: E = 120 V/-120°   Br. 9: R = 60 Ω
Br. 3: E = 120 V/120°   Br. 10: L = 56 mH
Br. 4: R = 10 kΩ   Br. 11: R = 21.2 Ω
Br. 5: R = 10 kΩ   Br. 12: C = 177 µF
Br. 6: R = 10 kΩ   Br. 13: R = 26 Ω
Br. 7: voltage reference   Br. 14: R = 0.01 Ω
                             Br. 15: R = 0.01 Ω

Record on the circuit diagram the current in each phase of the load, the line currents, and the currents in each phase of the source.

As there is no neutral connection, the phasor sum of the line currents must be zero. Sum the phasors of the three line currents.

$$I_A + I_B + I_C = \underline{\hspace{3cm}} \angle \underline{\hspace{2cm}}° \text{ A}$$

Comment on the comparison of the current in each phase of the source and the load.

Three-Phase DELTA-WYE Connected System

OBJECTIVE

To analyze a DELTA-WYE connected circuit for the load and source currents.

THEORY

Three-phase systems may be connected in the DELTA-WYE configuration. No neutral line is available with this connection. The load may be balanced or unbalanced and in either case the line currents must sum to zero. An unbalanced WYE connected load would cause the "floating neutral" condition and was analyzed in AC Exercise 1 so only the balanced load will be considered here.

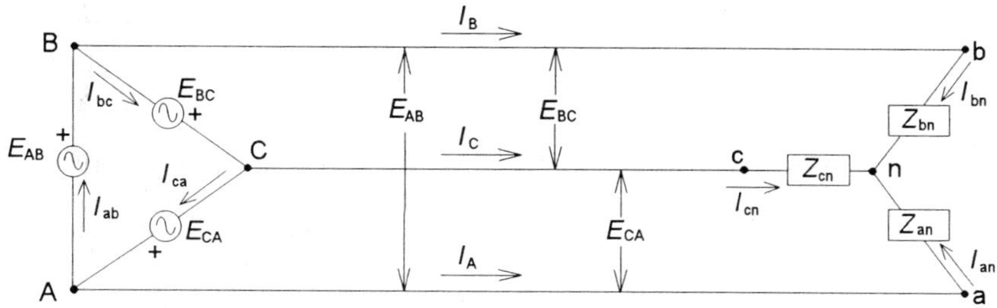

Figure 21.1

For a balanced load

$$V_{phase} = V_{line}/\sqrt{3} \text{ in magnitude but shifted } 30°$$

$$I_{line} = I_{phase} \text{ of the load}$$

Phase current of the **source** $= (1/\sqrt{3})\, I_{line}$

Power total $= 3P_{phase} = \sqrt{3}\,V_L I_L \cos\theta$ where θ is the load impedance angle

PROCEDURE

The circuit in Figure 21.2 represents a DELTA connected alternator or transformer supplying power to a balanced WYE-connected load.

Draw and solve the circuit for a frequency of 60 Hz.

Mark on the circuit diagram the phase currents of the load and the source.

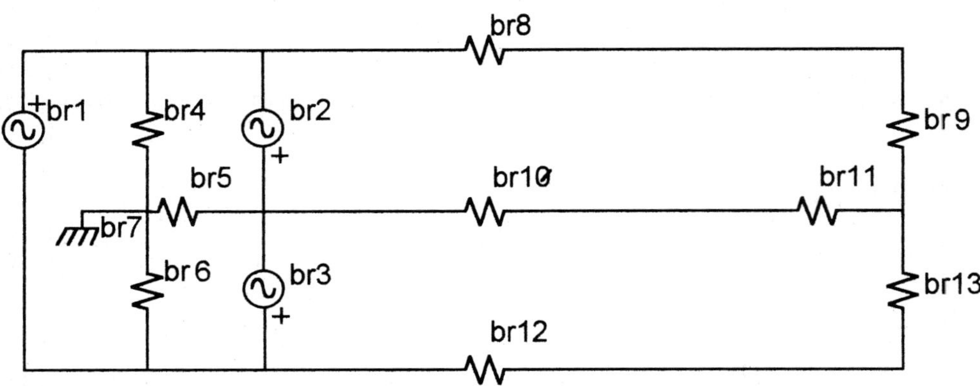

Figure 21.2

Net List for the Circuit

Br. 1: E = 120 V/0°   Br. 7: voltage reference
Br. 2: E = 120 V/-120°   Br. 8: R = 0.01 Ω
Br. 3: E = 120 V/120°   Br. 9: R = 10.4 Ω
Br. 4: R = 10 kΩ   Br. 10: R = 0.01 Ω
Br. 5: R = 10 kΩ   Br. 11: R = 10.4 Ω
Br. 6: R = 10 kΩ   Br. 12: R = 0.01 Ω
                                 Br. 13: R = 10.4 Ω

Calculate the magnitude of the phase current of the source as $I_{phase}/\sqrt{3}$ = _______ A.

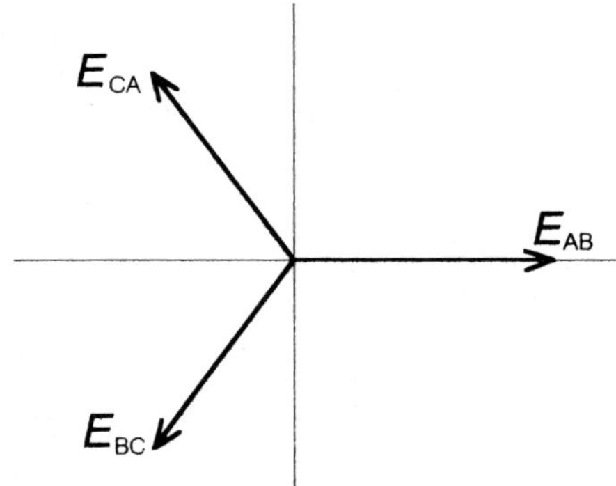

Phasor Diagram for
DELTA-WYE System

Using the solution results, draw the phase currents of the source and the line currents approximately to scale.

Comment on the magnitude and phase angles of the source currents.

Comment

Calculate the total power in the load as $3P_{phase} = 3V_{\mathrm{P}} I_{\mathrm{P}} \cos\theta$ = _____________ W.

The *Breadboard* result for the total power in the load is _____________ W.

Calculate the total power from the source as $P_{\mathrm{T}} = \sqrt{3} V_{\mathrm{P}} I_{\mathrm{P}} \cos\theta$ = ___________ W.

Two-Wattmeter Measurement of Three-Phase Power

OBJECTIVE

To confirm that the total power in a three-phase load can be measured using only two wattmeters.

THEORY

The total power of a three-phase system, whether balanced or unbalanced, can be measured using only two wattmeters whose current coils carry the line current in any two lines and whose voltage coils are connected to the remaining line. The total power is the algebraic sum of the two wattmeter readings. In circuits where the power factor of the load is less than 0.5 (cos(−60°)) or where the load is badly unbalanced, one of the wattmeters will read negatively and the wattmeter readings are subtracted.

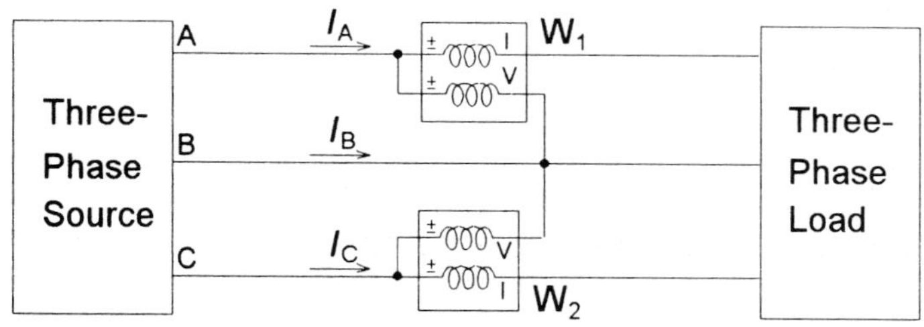

Figure 22.1

PROCEDURE Part 1

In the circuit shown in Figure 22.2, the current coils of the wattmeters are represented as low value resistors in branches 4 and 5. The voltage coils are the high value resistors of branches 6 and 7. An unbalanced load connected in DELTA to a WYE source has been chosen for this exercise although the two wattmeter method applies to any three-wire system.

On the circuit diagram of Figure 22.2, label the voltages and currents to correspond to those of Figure 22.1.

Draw the circuit shown in Figure 22.2 and solve for a frequency of 60 Hz.

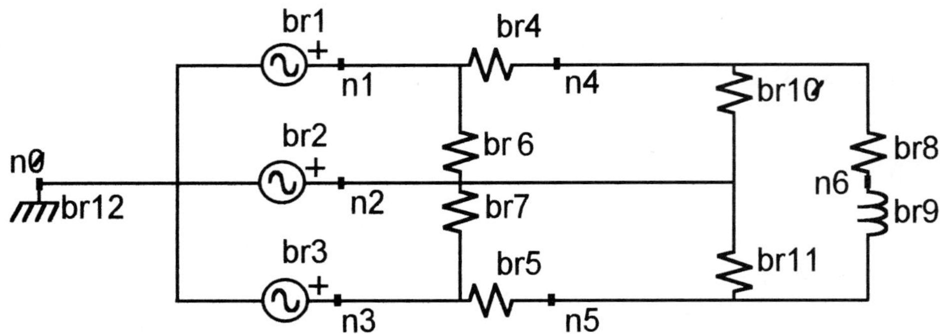

Figure 22.2

Net List for the Circuit

Br. 1: E = 120/0° V   Br. 7: R = 10 kΩ
Br. 2: E = 120/-120° V   Br. 8: R = 3 Ω
Br. 3: E = 120/120° V   Br. 9: L = 21.22 mH
Br. 4: R = 0.001 Ω   Br. 10: R = 20.8 Ω
Br. 5: R = 0.001 Ω   Br. 11: R = 10.4 Ω
Br. 6: R = 10 kΩ   Br. 12: voltage reference

Record the values of the current in the current coil of each wattmeter (branches 4 and 5) and the voltage across the voltage coils (branches 6 and 7) on the circuit diagram. Obtain the total power in the load from the solution results (branches 8, 10, and 11).

$$P_T = \underline{\hspace{4cm}} \text{ W}$$

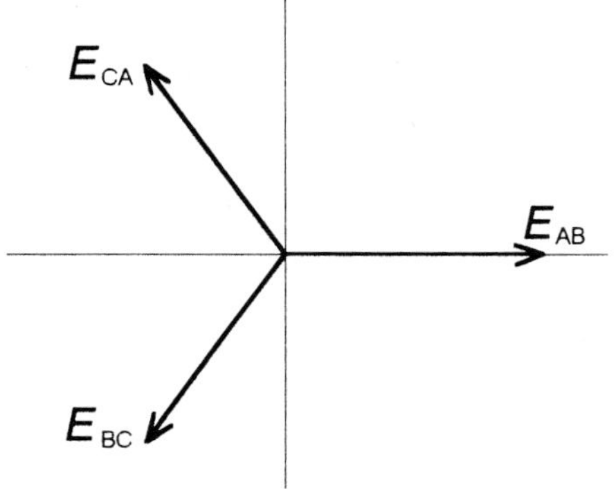

Phasor Diagram of
Wattmeter current and voltage

The readings of the wattmeters are calculated as VIcosθ where θ is the angle between the voltage and current as seen by the wattmeter. To determine the angle, a phasor diagram of the current and voltage of each wattmeter is required. Draw the phasor diagram for the current in branches 4 and 5, and the branch voltages of branches 6 and 7.

Wattmeter 1 reads $V_{AB}I_A\cos\theta$

$$I_{\text{branch 4}} = \underline{\hspace{3cm}} \angle \underline{\hspace{2cm}} °\ A$$
$$V_{\text{branch 6}} = \underline{\hspace{3cm}} \angle \underline{\hspace{2cm}} °\ V$$

The phase angle difference between V_{AB} and I_A is $\theta = \underline{\hspace{4cm}} °$

The reading for wattmeter 1 is $V_{\text{branch 6}}\,I_{\text{branch 4}}\,\cos\theta = \underline{\hspace{4cm}}$ W.

Wattmeter 2 reads $V_{CA}I_C\cos\phi$

$$I_{\text{branch 5}} = \underline{\hspace{3cm}} \angle \underline{\hspace{2cm}} °\ A$$
$$V_{\text{branch 7}} = \underline{\hspace{3cm}} \angle \underline{\hspace{2cm}} °\ V$$

Note that the voltage that wattmeter 2 measures is V_{CA} which is the reverse subscript from V_{AC}. On the phasor diagram the voltage of branch 7 must be reversed or rotated through 180°.

The phase angle difference between V_{CA} and I_C is $\phi = \underline{\hspace{4cm}} °$

The reading for wattmeter 2 is $V_{\text{branch 7}}\,I_{\text{branch 5}}\,\cos\phi = \underline{\hspace{4cm}}$ W.

The sum of the two wattmeter readings $= \underline{\hspace{4cm}}$ W.

Compare the result to the value of P_T obtained above $= \underline{\hspace{2cm}}$ W.

PROCEDURE Part 2

If the phase angle of the load impedances is 60 degrees or greater, one of the wattmeters will read negatively. In practice this is observed as a down scale reading and the voltage coil connections of the wattmeter would be reversed. The reading obtained would then be taken as negative.

On the circuit diagram of Figure 22.3 label the voltages and currents to correspond to those of Figure 22.1.

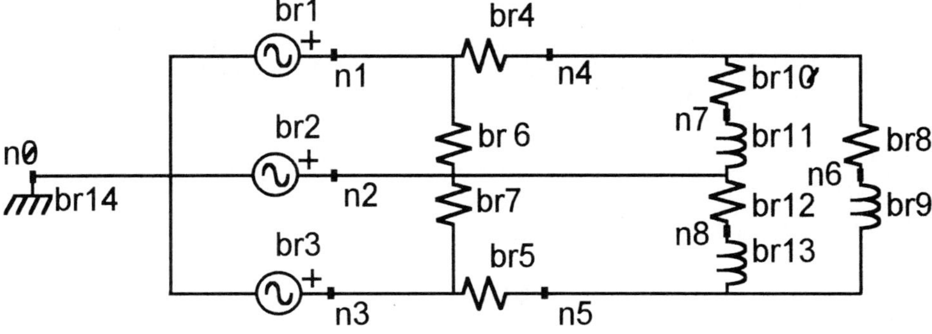

Figure 22.3

Net List for the Circuit

Br. 1: E = 120/0° V
Br. 2: E = 120/-120° V
Br. 3: E = 120/120° V
Br. 4: R = 0.001 Ω
Br. 5: R = 0.001 Ω
Br. 6: R = 10 kΩ
Br. 7: R = 10 kΩ

Br. 8: R = 7 Ω
Br. 9: L = 51.9 mH
Br. 10: R = 7 Ω
Br. 11: L = 51.9 mH
Br. 12: R = 7 Ω
Br. 13: L = 51.9 mH
Br. 14: voltage reference

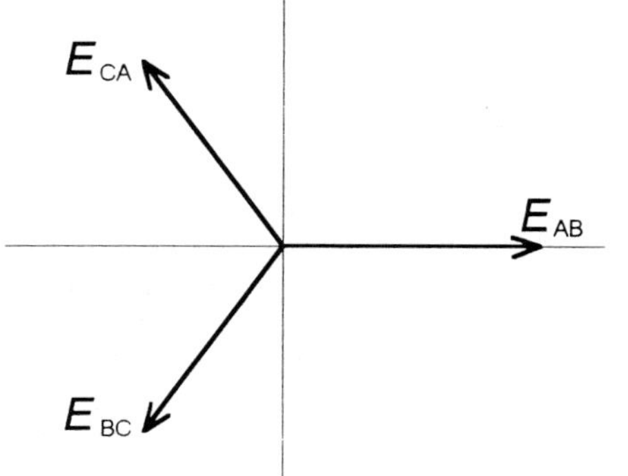

Phasor Diagram of
Wattmeter current and voltage

The readings of the wattmeters are calculated as VIcosθ, where θ is the angle between the voltage and current as seen by the wattmeter. To determine the angle, a phasor diagram of the current and voltage of each wattmeter is required. Draw the phasor diagram for the current in branches 4 and 5, and the branch voltages of branches 6 and 7.

Wattmeter 1 reads $V_{AB}I_{A}\cos\theta$

$$I_{\text{branch 4}} = \underline{\hspace{3cm}} \angle \underline{\hspace{2cm}} °\ A$$
$$V_{\text{branch 6}} = \underline{\hspace{3cm}} \angle \underline{\hspace{2cm}} °\ V$$

The phase angle difference between V_{AB} and I_A is $\theta° =$ _______________ °

The reading for wattmeter 1 is $V_{branch\ 6}\ I_{branch\ 4}\cos\theta =$ _______________ W.

Wattmeter 2 reads $V_{CA}I_{C}\cos\phi$

$$I_{\text{branch 5}} = \underline{\hspace{3cm}} \angle \underline{\hspace{2cm}} °\ A$$
$$V_{\text{branch 7}} = \underline{\hspace{3cm}} \angle \underline{\hspace{2cm}} °\ V$$

Note that the voltage that wattmeter 2 measures is V_{CA} which is the reverse subscript from V_{AC}. On the phasor diagram the voltage of branch 7 must be reversed or turned through 180°.

The phase angle difference between V_{CA} and I_C is $\phi =$ _______________ °

The reading for wattmeter 2 is $V_{branch\ 7}\ I_{branch\ 5}\cos\phi =$ _______________ W.

The sum of the two wattmeter readings = _______________ W.

Compare the result to the *Breadboard* solution of $P_T =$ _______________ W.

Power Transformers

OBJECTIVE

To determine the effect of different loads on the voltage output of a transformer.

THEORY

Transformers are used to step up or step down AC voltages and to provide isolation from the power source. The voltage transformation is directly proportional to the ratio of the number of turns of the primary and secondary winding of the transformer. The current transformation is inversely proportional to this ratio. The circuit of Figure 23.1 is the complete model of an iron core power transformer. The parallel resistor (R_o) and inductor (X_o) represent the core losses and magnetizing current of the transformer. These losses and the magnetizing current are normally small in comparison to the rating of the transformer and can be neglected in an analysis.

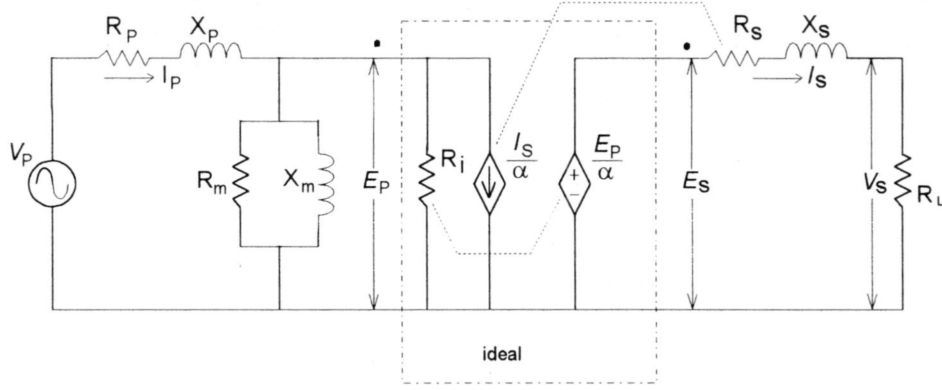

Figure 23.1

The transformer to be modeled is rated 60 kVA, 2400/240 volts with the winding resistance and inductance as shown in the net list. The model as shown in Figure 23.2 is constructed using a voltage-controlled voltage-source (VCVS in branch 7) to simulate the induced secondary voltage with a gain equal to the inverse transformation ratio ($1/\alpha = N_2/N_1 = E_S/E_P$). The controlling branch voltage is branch 4 (E_P), the counter-emf of the primary winding.

The current-controlled current-source (CCCS) of branch 5 in the primary represents the inverse current relationship of the transformer, given as $I_P = I_S(N_2/N_1) = I_S/\alpha$. The controlling branch for the CCCS is the current in the secondary winding (branch 8) of the transformer.

125

To achieve the dot polarity shown the VCVS is connected with the polarity indicated in Figure 23.2. Reversing the polarity of the VCVS will reverse the polarity of the secondary winding.

PROCEDURE Part 1

Of particular interest in analyzing the operation of a transformer is the voltage regulation under different load conditions. Voltage regulation is the change in output voltage of the transformer from no load to full load. The voltage change is due to the voltage drop across the internal resistance and reactance of the windings. The circuit of Figure 23.2 neglects the effect of exciting current as it is usually very small in comparison to the rated current of the transformer.

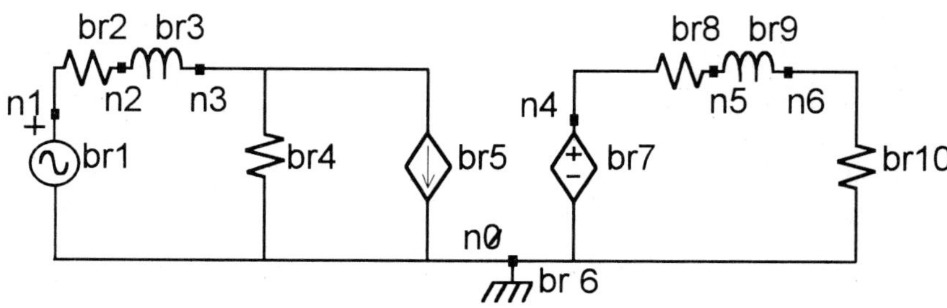

Figure 23.2

Net List for the Circuit

Br. 1: E = 2400 V/0°   Br. 7: VCVS Av = 0.102
Br. 2: R = 0.8 Ω   control branch 4
Br. 3: L = 2.4 mH   Br. 8: R = 0.008 Ω
Br. 4: R = 1 MΩ   Br. 9: L = 0.024 mH
Br. 5: CCCS hfe = 0.102   Br. 10: R = 0.982 Ω
     control branch 8
Br. 6: voltage reference

Unity Power Factor Load

Draw and solve the circuit of Figure 23.2 for a frequency of 60 Hz.

On the circuit diagram, mark the currents and the node voltages at nodes 1, 3, 4, and 6.

The full-load unity power factor output voltage ($V_{branch\ 6}$) is _______________ V.

The kVA output is ($I_{branch\ 6}V_{branch6}$)/1000 = _______________ kVA.

No Load

To obtain the no-load output voltage, a very high value resistor (a voltmeter) is used as the load. Replace branch 6 with a 1 MΩ resistor and solve for the output voltage.

The no-load output voltage is ______________ V.

The voltage regulation for a resistive load is calculated as

$$\text{VR} = [(V_{NL} - V_{FL})/V_{FL}] \cdot 100 = \underline{\hspace{3cm}} \%.$$

Leading Power Factor Load

Replace the load with a 0.502 Ω resistor and a 2500 μF capacitor in series. Solve for the output voltage.

The kVA output is $(I_{\text{branch 6}} V_{\text{branch 6}})/1000 = \underline{\hspace{4cm}}$ kVA.

The voltage output is ______________ V.

The voltage regulation for a lagging power factor load is ______________ %.

Lagging Power factor Load

Replace the load (branch 10) with a 0.624 Ω resistor and a 2.21 mH inductor in series. Solve for the output voltage.

The kVA output is $(I_{\text{branch 6}} V_{\text{branch 6}})/1000 = \underline{\hspace{4cm}}$ kVA.

The voltage output is ______________ V.

The voltage regulation is ______________ %.

Comment on the voltage regulation of the transformer for the three types of load.

Comment

PROCEDURE Part 2

The efficiency of a power transformer is approximately 95% or better. Under full load conditions the losses are mainly the I^2R losses in the resistance of the windings and the core losses due to magnetic hysteresis and eddy current losses. To represent the core losses, the resistance R_o as shown in Figure 23.1 is added to the model. The exciting current drawn by the transformer is obtained in the model by adding the reactance X_o.

$$R_o = 5760 \ \Omega \text{ and } X_o = 2.56 \text{ H}$$

Add these two components to the circuit as in Figure 23.1 and solve the circuit for 60 Hz. The efficiency may be calculated as

$$\text{efficiency} = \eta\% = (\text{power out})/(\text{power out} + \text{losses}) \cdot 100.$$

(a) Calculate the losses for the resistance load.

I^2R loss in the primary winding (branch 2) = _____________ W.

I^2R loss in the secondary winding (branch 8) = _____________ W.

Core loss (R_o branch) = _____________________ W.

Total losses = _______________ W.

efficiency = $\eta\%$ = _________________ %

(b) Repeat the above for the resistance and inductance load used in procedure 1.

Total losses = I^2R and core losses = _________________ W

efficiency = $\eta\%$ = _________________ %

(c) Repeat the above for the resistance and capacitance load used in procedure 1.

Total losses = I^2R and core losses = _________________ W

efficiency = $\eta\%$ = _________________ %

FURTHER ANALYSIS

Refer to a transformer problem in your textbook, model the transformer, and draw and solve the circuit using *Breadboard*.

Power Flow Studies

24

OBJECTIVE

To study the power delivered from an alternator connected to a power grid.

THEORY

Power generating stations are connected to a common grid for transmission and distribution to load centers. This provides improved reliability of the system should an alternator be taken off service for cost economies or maintenance.

When alternators are first synchronized and connected to the power grid they are floating on the line and do not supply power to the system. For the alternator to begin delivering power, the setting of the speed governor of the prime mover must be adjusted for increased speed of rotation. Since the synchronous speed of the alternator is fixed by the frequency of the system, the effect is to increase the power angle of the alternator and thus deliver power to the system.

Increasing the generated voltage of the alternator by increasing the DC current to its field poles does not cause the alternator to pick up load as in a DC system. Instead, an increase in generator voltage above that of the system will cause the alternator to deliver **lagging** reactive volt-amperes. The reactive component of current will cause a voltage drop due to armature reaction (equivalent to a voltage drop across the internal synchronous impedance of the alternator) and lower its terminal voltage to equal that of the system. If the generated voltage is decreased the alternator will provide **leading** reactive volt-amperes which will cause an increase in the terminal voltage to equal that of the system.

PROCEDURE Part 1

The circuit of Figure 24.1 simulates the alternators of the system (branch 5) connected to a grid and providing power to the load on the system represented as branch 8. Branches 6 and 7 represent the combined synchronous impedance of the alternators already on line. The incoming alternator is represented by the source of branch 1 and initially is floating (not providing power) on the line. Branches 2 and 3 represent its synchronous impedance.

Solve the circuit using *Breadboard* at a frequency of 60 Hz to obtain the voltages and currents of the network.

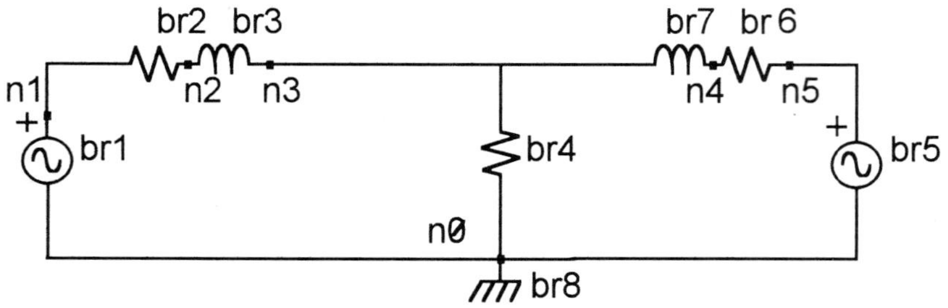

Figure 24.1

Net List for the Circuit

Br. 1: $E = 1200$ V/$0°$ Br. 5: $E = 1206.1$ V/$2.8°$
Br. 2: $R = 36$ m Ω Br. 6: $R = 0.36$ mΩ
Br. 3: $L = 1.36$ mH   Br. 7: $L = 13.6$ μH
Br. 4: $R = 0.1$ Ω Br. 8: voltage reference

Mark the branch currents and node voltages on Figure 24.1.

The load voltage ($V_{\text{branch 4}}$) is ______________ $\angle$ ______ ° kV.
The current in the load ($I_{\text{branch 4}}$) is ______________ $\angle$ ______ ° A.
The power in the load ($P_{\text{branch 4}}$) is ____________ MW.
The current delivered by the alternator ($I_{\text{branch 1}}$) is ____________ $\angle$ ______ ° A.
The power delivered by the alternator ($P_{\text{branch 1}}$) is ______________ kW.
The reactive volt-amperes delivered by the alternator ($Q_{\text{branch 1}}$) is ______ kVARS.
The power delivered by the system ($P_{\text{branch 5}}$) is ____________ MW.
The reactive volt-amperes delivered by the system ($Q_{\text{branch 5}}$) is ______ - ______ kVARS.
The alternator is delivering ($P_{\text{branch 1}}/P_{\text{branch 4}}$)·100____________ % of the load power.

PROCEDURE Part 2

To cause the alternator to pick up load its power angle must be advanced by setting the speed governor to a higher speed setting. The effect is to cause a leading phase angle of the voltage of the alternator. Edit the circuit diagram to change the voltage of the incoming alternator (branch 1) to $1200\angle+22°$ volts.

Solve the circuit for 60 Hz.

The line voltage ($V_{\text{branch 4}}$) is ______________ $\angle$ ______ ° kV.
The current in the load ($I_{\text{branch 4}}$) is ______________ $\angle$ ______ ° A.
The power in the load ($P_{\text{branch 4}}$) ____________ MW.
The current delivered by the alternator ($I_{\text{branch 1}}$) is ____________ $\angle$ ______ ° A.
The power delivered by the alternator ($P_{\text{branch 1}}$) is ______________ MW.
The reactive volt-amperes delivered by the alternator ($Q_{\text{branch 1}}$) is ______ kVARS.
The power delivered by the system ($P_{\text{branch 5}}$) is ____________ MW.
The reactive volt-amperes delivered by the system ($Q_{\text{branch 5}}$) is ______ kVARS.
The alternator has picked up ($P_{\text{branch 1}}/P_{\text{branch 4}}$)·100 ______ % of the load power.

Draw the power triangles for the incoming alternator (branch 1) and the system alternators (branch 5).

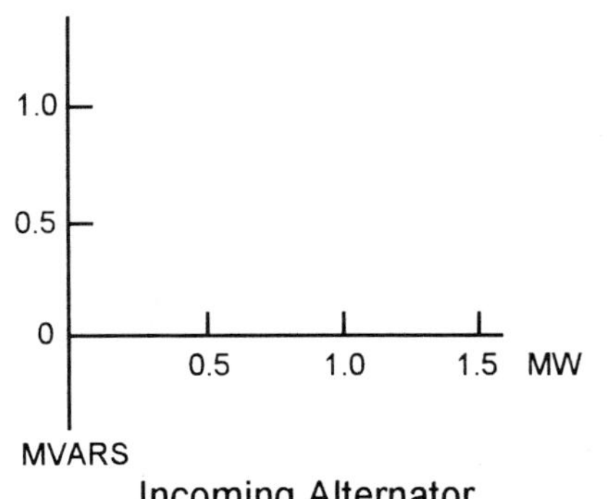
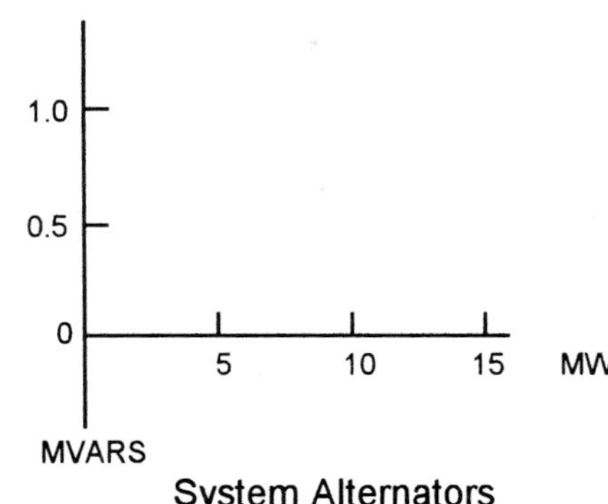

Incoming Alternator   System Alternators

PROCEDURE Part 3

If the voltage of the alternator (branch 1) is increased and the power angle is set back to zero, the alternator will deliver reactive volt-amperes lagging in order to cause the terminal voltage to equal that of the system. Edit the circuit to change the voltage of the alternator to $1500\angle 0°$ volts. Solve the circuit for 60 Hz.

The line voltage ($V_{branch\ 4}$) is ______________ $\angle$ _______ ° kV.
The current in the load ($I_{branch\ 4}$) is _____________ $\angle$ _______ ° A.
The power delivered to the load ($P_{branch\ 4}$) is _____________ MW.

The current delivered by the alternator ($I_{branch\ 1}$) is ____________ $\angle$ _______ ° A.
The power delivered by the alternator ($P_{branch\ 1}$) is _______________ kW.
The reactive volt-amperes delivered by the alternator is ($Q_{branch\ 1}$) _______ kVARS.

The power delivered by the system ($P_{branch\ 5}$) is _______________ MW.
The reactive volt-amperes delivered by the system ($Q_{branch\ 5}$) is ________ kVARS.

The alternator has picked up ($P_{branch\ 1}/P_{branch\ 4}$)·100 _______ % of the load power.

Draw the power triangles for the incoming alternator (branch 1) and the system alternators (branch 5).

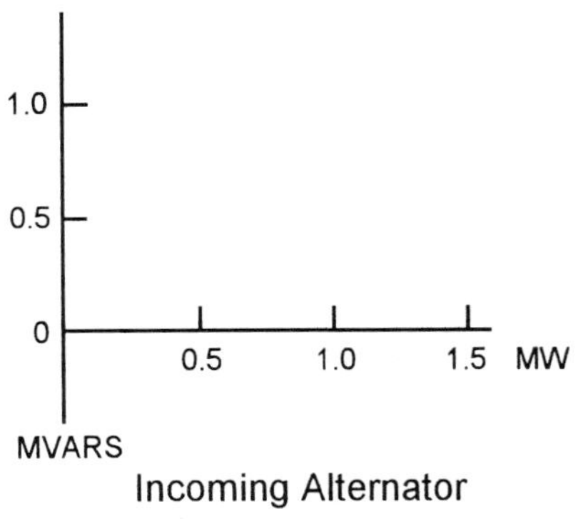
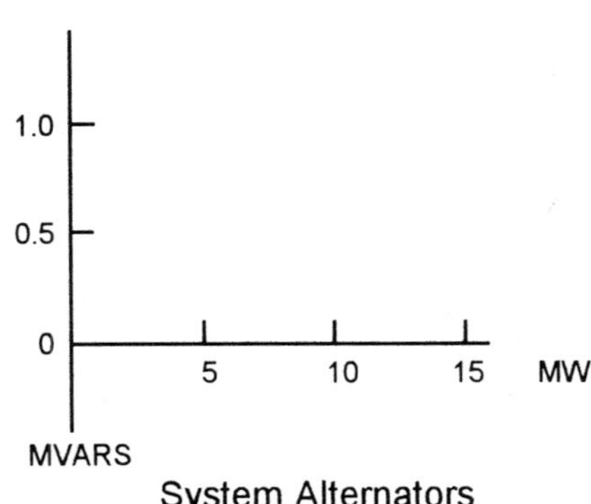

Incoming Alternator   System Alternators

PROCEDURE Part 4

To observe the effect of lowering the generated voltage of the alternator, edit the circuit and change the value of the incoming alternator voltage (branch 1) to $900\angle 1.1°$ volts. Solve the circuit for 60 Hz.

The line voltage ($V_{\text{branch 4}}$) is ________________ $\angle$ ________ ° kV.
The current in the load ($I_{\text{branch 4}}$) is ________________ $\angle$ ________ ° A.
The power delivered to the load ($P_{\text{branch 4}}$) ________________ MW.

The current delivered by the alternator ($I_{\text{branch 1}}$) is ________ $\angle$ ________ ° A.
The power delivered by the alternator ($P_{\text{branch 1}}$) is ________________ MW.
The reactive volt-amperes **received** by the alternator is ($Q_{\text{branch 1}}$) ________ kVARS.

The power delivered by the system $P_{\text{branch 5}}$ is ________________ MW.
The reactive volt-amperes delivered by the system ($Q_{\text{branch 5}}$) is ________ kVARS.

The alternator has picked up ($P_{\text{branch 1}}/P_{\text{branch 4}}$)·100 ________ % of the load power.

Draw the power triangles for the incoming alternator (branch 1) and the system alternators (branch 5).

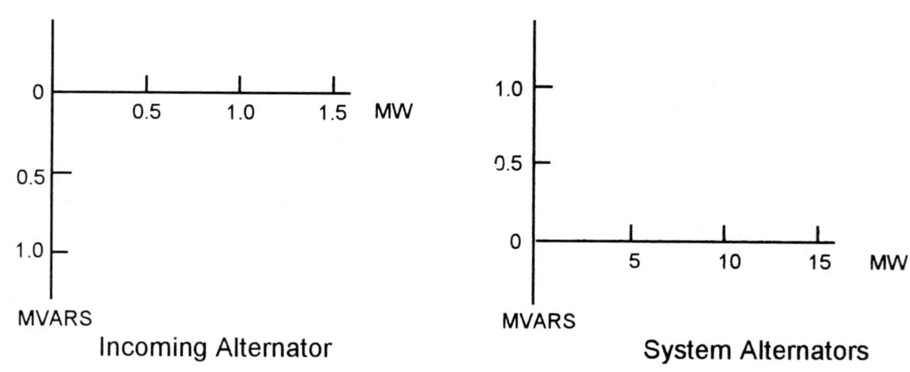

Write a summary for the three conditions of the incoming alternator.

Summary

FURTHER ANALYSIS

If the power angle of the alternator (branch 1) is decreased to represent a shedding of load by its prime mover, the alternator will act as a synchronous motor and absorb power. Change the voltage of the alternator (branch 1) to $1200\angle -10°$ and analyze the solution as above.

BJT Transistor DC Model and Base-Bias Circuit

OBJECTIVE

To model a BJT in its DC mode of operation and to study the effect of beta (β) on the quiescent operating point of a base-biased (fixed-biased) transistor.

THEORY

The many effects that take place in a semiconductor device operating in a circuit make the complete model of the device very complicated. The electrical characteristics of devices are nonlinear, and their parameters are dependent on the temperature and voltage applied to them.

In order to readily analyze transistor circuits, approximate models are used that neglect the smaller effects, and the analysis is restricted to the linear regions of the characteristic curves. Considering the large variations of device parameters as given in manufacturer's data sheets, the approximations are usually adequate for most applications. The software program *Breadboard* provides linear, steady-state analysis of circuits, and the user must recognize that the device is considered to be operating in the linear region of its characteristic curves.

Figure 1.1 shows a simplified DC transistor model in which the smaller effects of its operation have been eliminated. The components shown on the figure and their approximate values are given below.

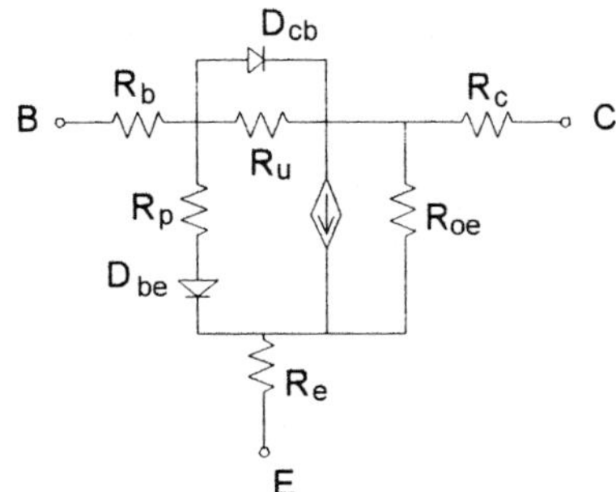

R_b – base spreading resistance (10 Ω)
R_u – collector-to-base leakage (1 GΩ)
R_p – base emitter diode resistance (100 Ω)
R_{oe} – output resistance (>1 MΩ)
D_{cb} – reverse-biased collector-to-base diode
D_{be} – forward biased base-emitter diode

Simplified DC model
Figure 1.1

"""

The ideal DC model shown in Figure 1.2 eliminates the bulk resistance of the base, collector, and emitter and the spreading resistance of the base, because these resistances are very small compared to the resistances that will be added in order to bias the device for proper operation. The voltage source V_{BE} represents the base-emitter forward-bias diode voltage and has a value of 0.3 V for germanium and 0.7 V for silicon transistors. The DC current gain of the transistor (beta) is defined as $\beta = I_C/I_B$, and given as h_{FE} on manufacturers' specification sheets. The current-controlled current source (CCCS) is shown causing a current (I_C) of β times the base current (I_B) to flow from the collector to the emitter.

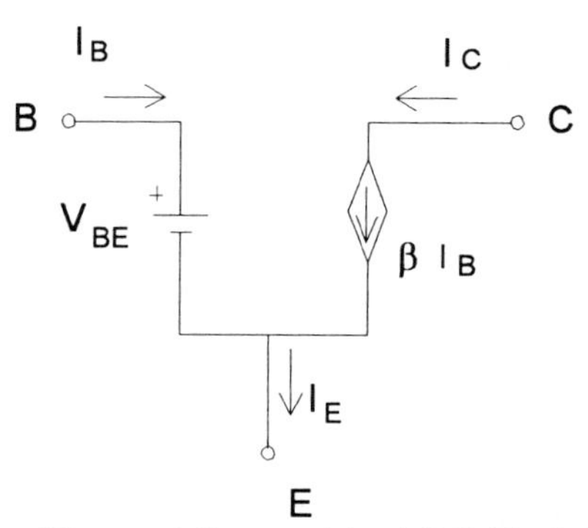

Figure 1.2   Ideal DC Model

A bipolar junction transistor may be considered to be a current amplifying device, where the current in the base of the transistor determines the current that flows in the collector of the device. The current that flows in the collector is β (the DC current amplification) times as large.

With base-bias, the base current of a transistor is held constant and approximately equals V_{CC}/R_B . The operating point will then depend on the current gain of the transistor, and due to manufacturing tolerances the operating point will vary for each transistor used in the circuit. The current gain also varies with temperature (increasing as the temperature increases), moving the quiescent operating point, and may cause thermal runaway resulting in transistor burnout. Although base-bias is simple and requires only one resistor, the application of this type of bias in linear amplifiers is limited and is mainly used in switching circuits.

PROCEDURE

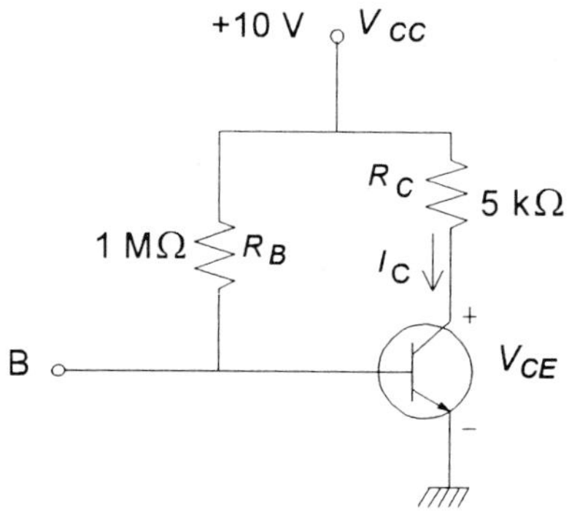

The circuit of Figure 1.3 is to be analyzed for four different values of current gain of the transistor, and the quiescent operating point is to be plotted on the DC load line of Figure 1.4.

Figure 1.3

Using *Breadboard* in the DC analysis mode, draw the circuit model of Figure 1.4 for the values of components shown in the net list.

Note: Components should be drawn in the sequence that follows the conventional current flow in them.

On Figures 1.3 and 1.4, mark the current through and the voltage across each component for the solution values obtained for $\beta = 100$. Study the relative magnitudes of the currents in each branch. The voltage across the CCCS is the V_{CE} of the transistor and the power dissipated in the transistor is $V_{CE}I_C$.

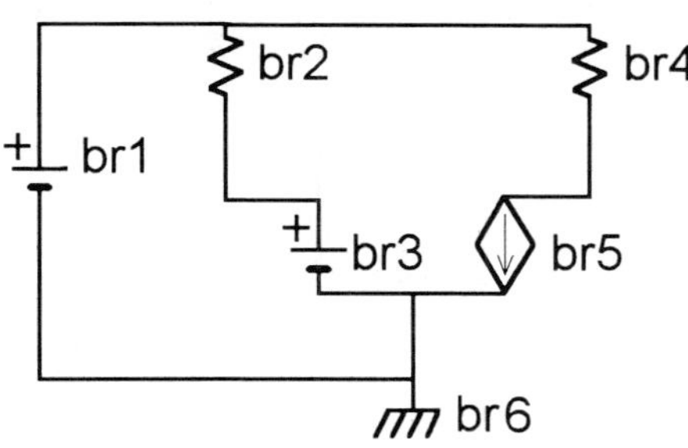

Figure 1.4

Net List For the Circuit

Br. 1: E = 10 V/0°
Br. 2: R = 1 MΩ
Br. 3: E = 0.7 V
Br. 4: R = 5 kΩ
Br. 5: CCCS, beta = 100
      control branch 2
Br. 6: voltage reference

Solve the circuit for the values of β shown in Table 1.1 and complete the table. From the results, plot the Q points on the load line in Figure 1.5.

Table 1.1

Beta (h_{FE})	I_B (branch 2)	I_C (branch 4)	V_{CE} (branch 5)
50			
100			
150			
200			

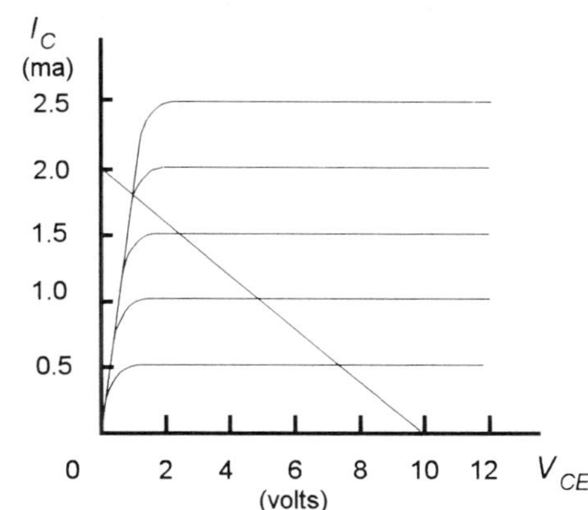

Figure 1.5

The operating point for the base-bias circuit can be calculated as

$$V_{RB} = V_{CC} - V_{BE}$$

where $V_{BE} = 0.7$ volts for silicon and 0.3 for germanium transistors.

$$I_B = V_{RB}/R_B \qquad \text{and} \qquad I_C = \beta I_B$$

Then $V_{CE} = V_{CC} - I_C R_C$

For a current gain of 100 calculate I_C and V_{CE} using the formulas above.

Does the calculated value of I_C and V_{CE} agree with the *Breadboard* results for a current gain of 100? _________

For $\beta = 100$, the power dissipated in the transistor is:

$$P = I_C V_{CE} = \underline{\hspace{2cm}} \text{ mW}$$

Study the results in the table and comment on the change of operating point as the current gain of the transistor was changed.

Comment

FURTHER ANALYSIS

Model the transistor that you will use in the laboratory for the experiment on the base-bias of a transistor and solve the circuit using *Breadboard*. Compare the results to those obtained in the laboratory.

Refer to your text on the topic of designing a base-bias circuit for a BJT transistor. Design the circuit. Model the circuit and use *Breadboard* to determine that the design is correct.

Collector-Feedback Bias

OBJECTIVE

To solve for the currents and voltages of a collector-feedback bias circuit, and to observe the stability of the quiescent operating point.

THEORY

The actual beta (h_{FE}) of a specific type of transistor will range between -50% and 150% due to manufacturing tolerance. Also, the operating temperature of the device affects the value of beta. Changes in beta will affect the quiescent (Q) operating point of the device.

In collector-feedback bias the base resistor is connected to the collector of the transistor rather than the power supply. If the collector current should increase due to a change in beta, the larger collector current will cause an increased voltage drop across the collector resistor R_C. This increased voltage drop will lower the voltage to the base resistor R_B, which will decrease the base current. This in turn will reduce the collector current that depends on the base current. This feedback effect stabilizes the collector current against changes in beta and tends to hold the quiescent operating point of the transistor fixed.

Note: Should the DC analysis show V_{CE} to be less than a few tenths of a volt, the operating point of the transistor would be in the saturation region of the characteristic curves, which is an improper bias condition for an amplifier circuit. Also, if V_{CE} at the operating point exceeds the maximum voltage rating for the transistor, a burnout condition would exist.

The following formulas are derived using Kirchhoff's voltage law for the circuit shown in Figure 2.1.

$$V_{CC} = V_{BE} + I_B R_B + (I_B + I_C)R_C$$

Since $\beta I_B = I_C$ then $(I_B + I_C)$ is equal to $I_B(1+\beta)$

and $V_{CC} = V_{BE} + I_B [(1+ \beta)R_C + R_B]$.

Separating I_B gives $I_B = (V_{CC} - V_{BE})/[(1 + \beta)R_C + R_B]$

From I_B above, $I_C = \beta I_B$

from which V_{CE} can be obtained as $V_{CE} = V_{CC} - I_C R_C$.

PROCEDURE

The circuit for collector-feedback bias is shown in Figure 2.1, and Figure 2.2 is the equivalent model.

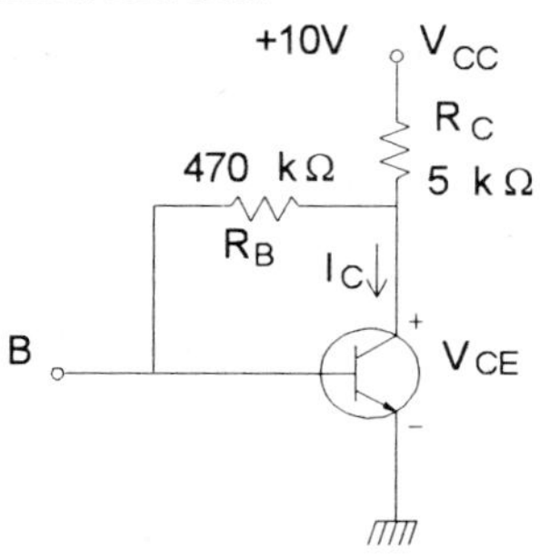

Figure 2.1

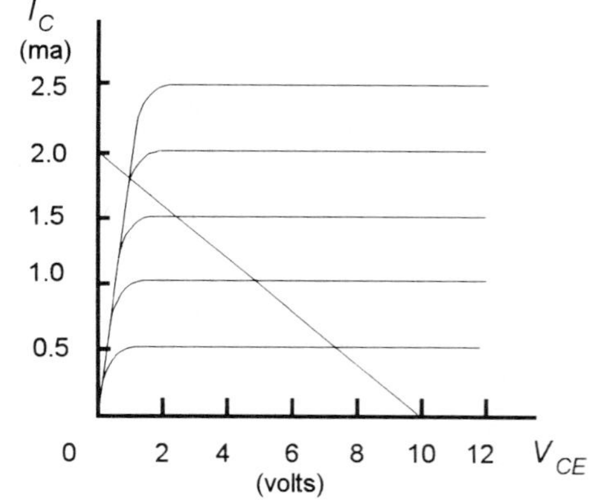

Figure 2.2

Label the base, emitter, and collector of the transistor in Figure 2.2. Using *Breadboard* in DC analysis mode, draw the circuit for the values of components in the net list and solve the circuit. Record the results in Table 2.1.

Net List For the Circuit

Br. 1: E = 10 V/0°   Br. 4: R = 470 kΩ
Br. 2: R = 5 kΩ   Br. 5: E = 0.7 V
Br. 3: CCCS, beta = 100   Br. 6: voltage reference
    control branch 4

Table 2.1

Beta (h_{FE})	I_B (branch 4)	I_C (branch 2)	V_{CE} (branch 3)
50			
100			
150			
200			

Figure 2.3

On Figures 2.1 and 2.2, mark the current through and the voltage across each component. Use KCL and KVL to confirm the solution is correct. For the remaining values of beta given in Table 2.1, edit the drawing, solve, and complete the table.

On the load line for the transistor (Figure 2.3), mark the operating point for each value of beta. Compare the stability of the operating point to those obtained in Electronics Exercise 1 for the transistor with base-bias.

Voltage-Divider Bias

OBJECTIVE

To investigate the stabilization of the quiescent operating point of a transistor by using voltage-divider bias.

THEORY

Voltage-divider bias is commonly used to overcome the instability of the transistor Q-point due to variations in current gain caused by manufacturing tolerance, operating temperature of the device, and the effect of changes in V_{BE} with temperature.

With reference to Figure 3.1, the voltage across the emitter resistor R_E is much larger than the voltage V_{BE} (approximately 0.7 V for silicon) and swamps any changes in V_{BE} due to temperature.

The operating point stabilization is achieved in the following manner. The voltage at the base of the transistor is held fixed by the voltage divider made up of resistors R_1 and R_2. If a transistor with a higher current gain is substituted, the emitter current would increase, causing a higher voltage drop at the emitter. Since the base current depends on the voltage difference between the base and emitter, the base current would decrease, reducing the tendency for the emitter current to increase and offsetting the effect of increased current gain.

The formulas for calculating I_C and V_{CE} are given below.

$$V_B = \frac{R_1}{R_1 + R_2} V_{CC} \qquad I_E = \frac{V_E}{R_E}$$

$$V_E = V_B - V_{BE} \qquad V_{CE} = V_{CC} - I_C R_C - I_E R_E$$

$$\text{asumming } I_E = I_C \qquad V_{CE} = V_{CC} - I_E(R_C + R_E)$$

PROCEDURE

Figure 3.1 is a circuit schematic for a voltage-divider bias circuit and Figure 3.3 is the equivalent circuit model. Figure 3.2 is the load line for the circuit. Note that a small base resistor has been added in order to provide a control branch resistor for the CCCS. It may be considered to be the bulk resistance of the base.

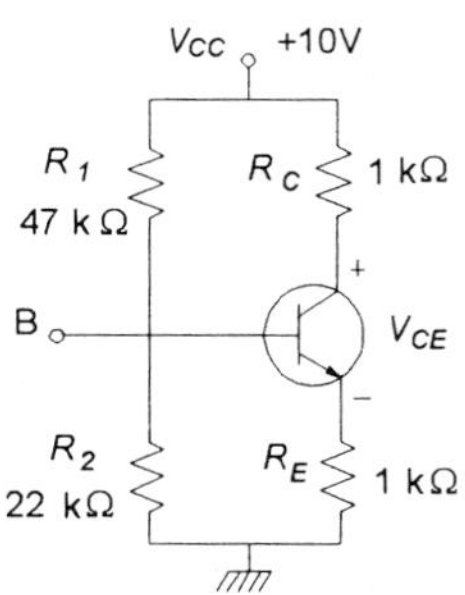

Figure 3.1

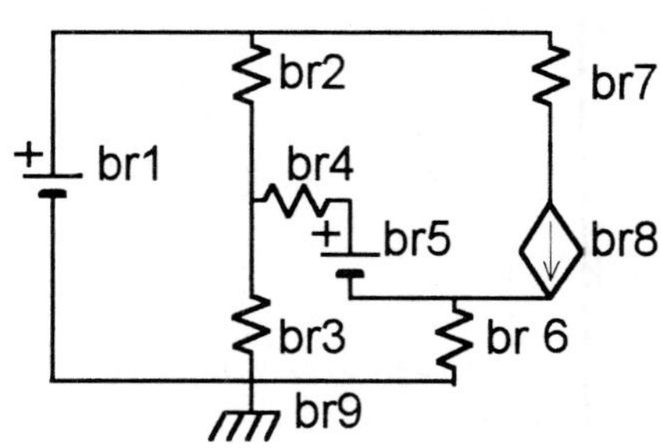

Figure 3.2

Mark the base, collector, and emitter on the circuit diagram of Figure 3.2. For the values of components given in the net list, draw and solve the circuit model using *Breadboard* in the DC analysis mode.

Net List For the Circuit

Br. 1: E = 10 V   Br. 5: E = 0.7 V   Br. 8: CCCS, beta = 100
Br. 2: R = 47 kΩ   Br. 6: R = 1.0 kΩ   control branch 4
Br. 3: R = 22 kΩ   Br. 7: R = 1.0 kΩ   Br. 9: voltage reference
Br. 4: R = 50 Ω

From the solution, mark on Figures 3.1 and 3.2 the current and voltage obtained for each component. Use KVL and KCL to confirm that the circuit voltages and currents are correct.

Modify the circuit and obtain solutions for values of beta of 50, 150, and 200. Complete the table for these values of beta.

Table 3.1

Beta	I_B	I_C	V_{CE}
(h_{FE})	(branch 4)	(branch 7)	(branch 8)
50			
100			
150			
200			

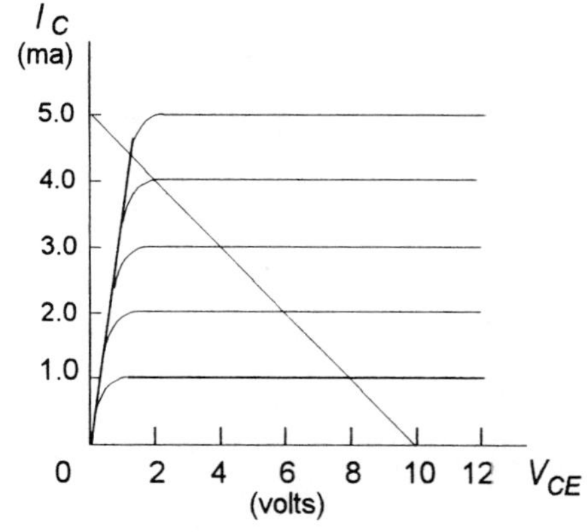

Figure 3.3

Comment on the comparative change in the operating points for voltage-divider bias with the changes observed in collector–feedback bias.

Emitter Bias

OBJECTIVE

To study the stability of the Q-point for an emitter-bias circuit.

THEORY

Emitter-bias is another form of biasing a transistor. It is used if a split-voltage power supply is available to the designer. The general problem to be overcome with transistor biasing is the instability of the Q-point due to variation of beta (h_{FE}) because of manufacturing tolerances and changes in operating temperature of the device. In a properly designed emitter-bias circuit, the base of the transistor is held at approximately zero volts and the emitter voltage therefore is -0.7 V. Since the emitter resistor voltage is held fixed by the $-V_{EE}$ supply voltage, the emitter voltage is almost independent of the value of h_{FE}.

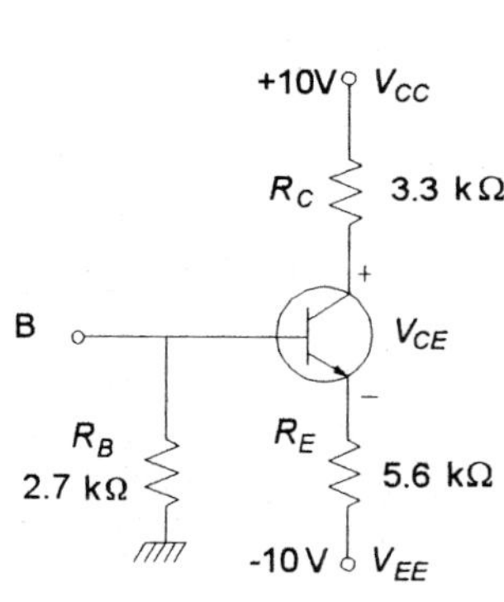

Figure 4.1

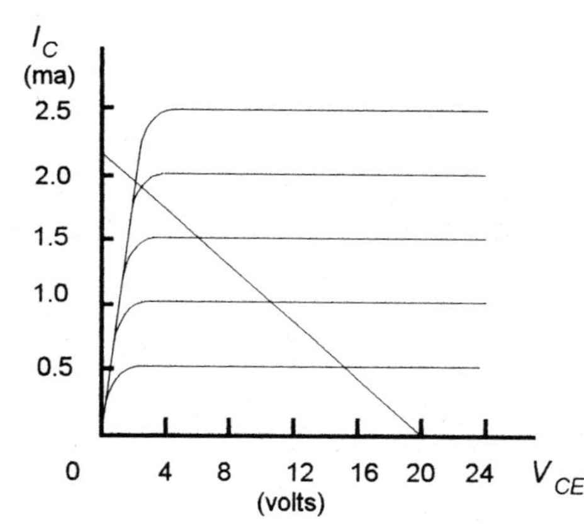

Figure 4.2

The formulas for calculating I_C and V_{CE} are given below.

$$I_E = \frac{V_{EE} - V_{BE}}{R_E + R_B / \beta_{dc}} \cong I_C$$

$$V_C = V_{CC} - I_C R_C \qquad V_E = I_E R_E - V_{EE}$$

$$V_{CE} = V_C - V_E$$

PROCEDURE

Figure 4.1 is the circuit schematic for a split-supply emitter-bias circuit and Figure 4.2 is the load line for the circuit. Draw the circuit model shown in Figure 4.3 using *Breadboard* in the DC mode for the values of components given in the net list and solve the circuit.

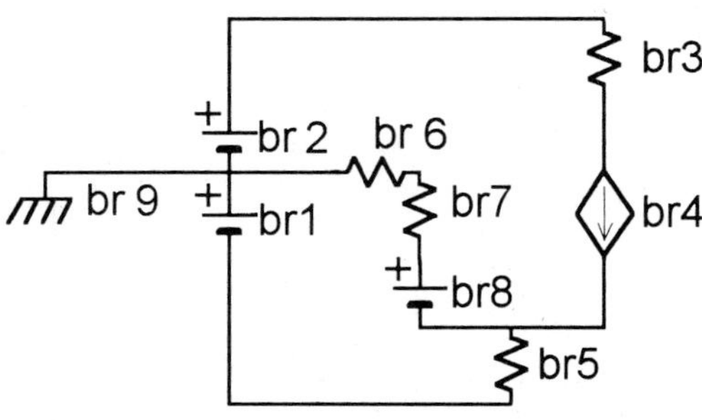

Figure 4.3

Net List for the Circuit

Br. 1: E = 10 V
Br. 2: E = 10 V
Br. 3: R = 3.3 kΩ
Br. 4: CCCS, beta = 100
          control branch 7
Br. 5: R = 5.6 kΩ
Br. 6: R = 2.7 kΩ
Br. 7: R = 10 Ω
Br. 8: E = 0.7 V
Br. 9: voltage reference

From the solution results, mark the branch currents on the circuit diagrams of Figures 4.1 and 4.3. Study the solution values for the circuit.

Enter the values for I_B, I_C, and V_{CE} in table 4.1

Table 4.1

Beta (h_{FE})	I_B (branch 7)	I_C (branch 3)	V_{CE} (branch 4)
50			
100			
150			
300			

Edit the *Breadboard* circuit to change the current gain of the CCCS to the remaining values of current gain in the table. Solve the circuit, mark the operating point on the load line of Figure 4.3, and label the points with the corresponding h_{FE} value.

Comment on the change in operating point with changes in h_{FE} and the resulting changes in base current and collector-emitter voltages.

Comment

Common–Emitter Amplifier

OBJECTIVE

To model a BJT common-emitter small-signal amplifier and to study its characteristics.

THEORY

In order to facilitate analysis of transistor amplifier circuits, an equivalent model is derived. The circuit in Figure 5.2 is the approximate h-parameter model of a transistor shown in Figure 5.1.

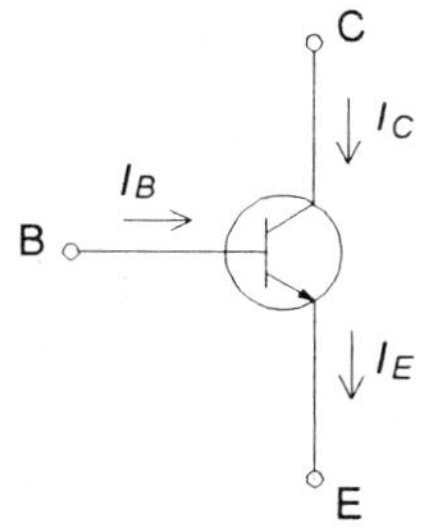

Figure 5.1

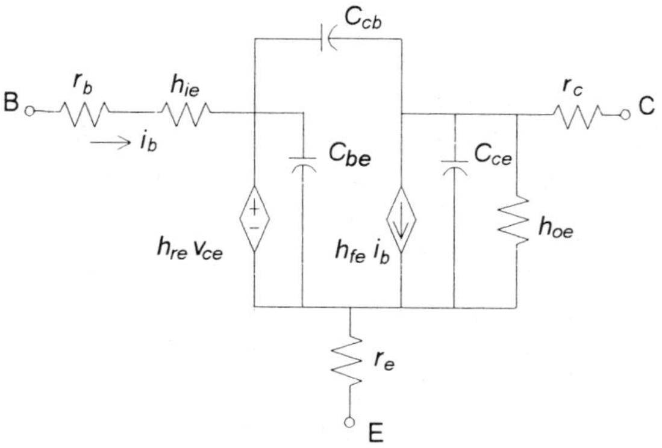

Figure 5.2

In this hybrid parameter model, the bulk resistance of the emitter, collector, and base regions are considered to be negligible. The junction-diode capacitances are very small and can be omitted if the operating frequency is below the cutoff of the transistor (<100 MHz). The effect of depletion layer-width changes (reverse voltage gain), represented as $h_{re}v_{ce}$, is also small (μvolts) and can be omitted from the model. The output admittance of the transistor h_{oe}, representing a large resistance (>100 kΩ), will be paralleled by a much lower value collector resistor (R_C) and may be removed from the model. The short-circuit input impedance (h_{ie}) is obtained from manufacturer's data sheets. The small-signal model then simplifies to that shown in Figure 5.3. *Breadboard* assumes the device is working on the linear, active region of the characteristic curves.

Another approximate model, the r'_e parameter model using the concept of the emitter diode AC resistance is shown in Figure 5.4. This model is equivalent to the h-parameter model since $r'_e = h_{ie}/h_{fe}$. The AC emitter diode resistance is defined as $\delta V_{BE}/\delta I_{EQ}$ (or approximately 25 mV/I_{EQ}), where I_{EQ} is the DC quiescent emitter current.

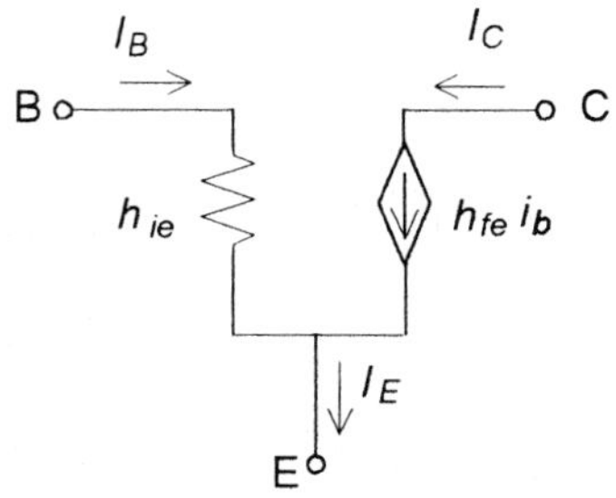

Figure 5.4 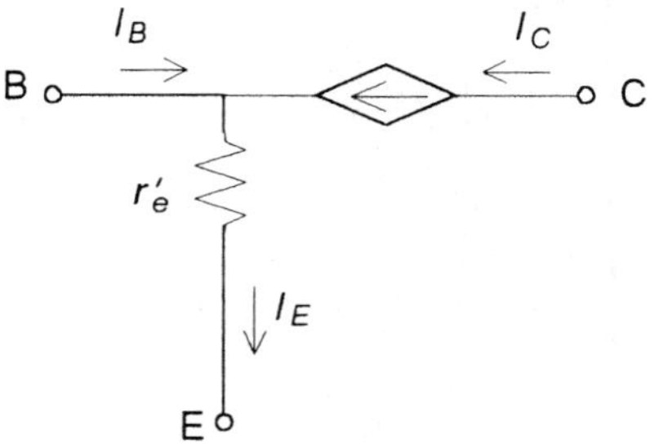   Figure 5.4

To analyze the bypassed common-emitter transistor amplifier shown in Figure 5.5, the DC sources are considered low-resistance paths for AC signals and are shorted and connected to the common line of the circuit. The AC-equivalent circuit is shown in Figure 5.6.

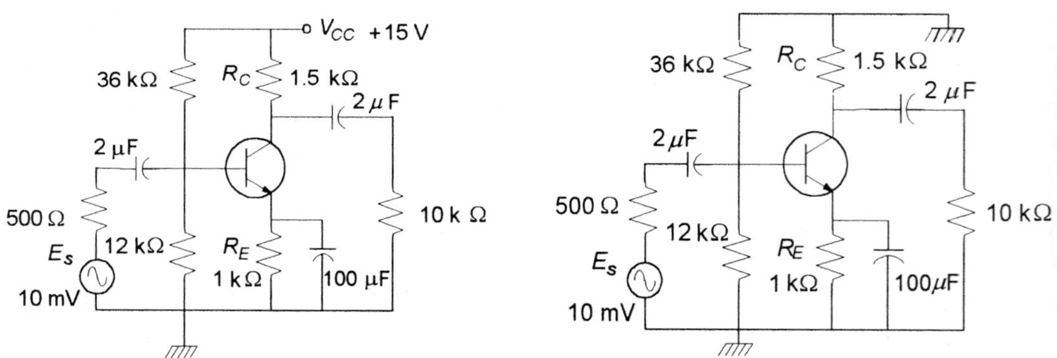

Figure 5.5   Figure 5.6

If the coupling capacitors (the 2μF capacitors) are of sufficiently large value that they do not affect the signal at the operating frequency of the transistor, they may be removed from the AC-equivalent circuit. Also, the emitter bypass capacitor (the 100 μF capacitor) effectively acts as a short circuit for AC across R_E and the emitter may be connected directly to the common line. The capacitors have been included in the model for this exercise to emphasize that their effect is negligible.

The hybrid parameters are obtained from the manufacturers' specification sheet for the type of transistor used. The minimum and maximum that can be expected are given for several values of collector current, and some interpolation is required to obtain a value to use. This is all the more reason the approximate model for the transistor is adequate.

The common-emitter amplifier provides the best voltage gain and current gain compared to the common-collector or common-base configurations. In addition the circuit provides comparatively high input impedance and a moderate output impedance (approximately equal to R_C), which are desired characteristics. A totally bypassed emitter causes some instability in voltage gain due to the dependence of r'_E on the operating temperature of the device. The loading of the amplifier will affect the voltage gain. The difference in voltage gain for a high- and moderate-value load resistor will be measured.

PROCEDURE Part 1

The hybrid parameters for the transistor in Figure 5.5 as obtained from the manufacturer's specification sheet for a 2N3904 transistor are h_{ie} = 3.5 kΩ, h_{fe} = 120, h_{re} = 1.3(10^{-4}), and h_{oe} = 8.5 μS.

Draw the circuit model of Figure 5.7 using the AC analysis option of *Breadboard* for the values shown in the net list. Obtain a solution for a frequency of 1.0 kHz. On the circuit diagram of Figure 5.5, mark the current through each component and the voltage at each node.

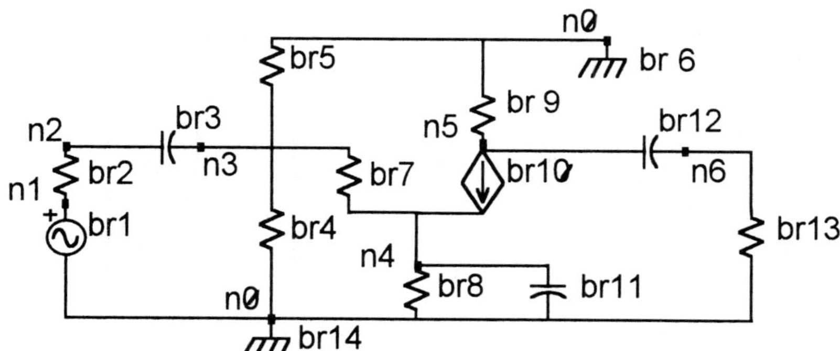

Figure 5.7

Net List For the Circuit

Br. 1: E = 10 mV/0°	Br. 9: R = 1.5 kΩ
Br. 2: R = 500 Ω	Br. 10: CCCS, beta = 120
Br. 3: C = 2.0 μF	control branch 7
Br. 4: R = 12 kΩ	Br. 11: C = 100 μF
Br. 5: R = 36 kΩ	Br. 12: C = 2.0 μF
Br. 6: voltage reference	Br. 13: R = 10 kΩ
Br. 7: R = 3.5 kΩ	Br. 14: voltage reference
Br. 8: R = 1.0 kΩ	

Calculate the voltage gain of the amplifier.

$$A_v = (V_{out}/V_{in}) = (V_{node\ 6})/(V_{node\ 2}) = \underline{\hspace{2cm}}$$

Calculate the current gain of the amplifier (I_{out}/I_{in}).

$$A_i = (I_{branch\ 13})/(I_{branch\ 2}) = \underline{\hspace{2cm}}$$

Calculate the input impedance of the amplifier. $Z_{in} = (V_{node\ 2})/(I_{branch\ 2}) = \underline{\hspace{1.5cm}}\ \Omega$

What is the power gain of the amplifier? $A_p = A_v A_i = \underline{\hspace{2cm}}$

What are the voltage drops across the coupling capacitors?

$$V_{branch\ 3} = \underline{\hspace{1.5cm}}\ V \qquad\qquad V_{branch\ 12} = \underline{\hspace{1.5cm}}\ V$$

What is the voltage across the emitter resistance and bypass capacitor?

$$V_{branch\ 11} = \underline{\hspace{1.5cm}}\ V$$

In the AC model, could the emitter have been connected to the common line without affecting the results? _________________

PROCEDURE Part 2

What is the open-output voltage gain of the amplifier? To obtain the value of open-output voltage, replace the load resistor in the *Breadboard* circuit with a very high value resistor such as $10^8\ \Omega$ and solve for a frequency of 1.0 kHz.

$$\text{open-output voltage} = E_{oc} = \underline{\hspace{2cm}}\ V$$

$$\text{unloaded voltage gain} = (V_{node\ 6})/(V_{node\ 2})\ \underline{\hspace{1.5cm}}$$

Compare the voltage gain of the loaded amplifier (part 1) to the unloaded voltage gain. ____

The output impedance of the amplifier can be calculated by Thevenin's theorem ($Z_{TH} = E_{oc}/I_{sc}$). Replace the load resistor (branch 13) with a very small value resistor (representing a short circuit) of the order of $0.0001\ \Omega$. Solve at the frequency of 1 kHz and obtain the current through this resistor.

$$Z_{out} = E_{oc}/I_{sc} = \underline{\hspace{2cm}}\ \Omega$$

FURTHER ANALYSIS

Edit the circuit drawing and replace the transistor with a substitute having $h_{fe} = 200$ and $h_{ie} = 1.0\ k\Omega$. Solve for a frequency of 1.0 kHz and determine the gain and input impedance as above. Comment on the stability of the amplifier.

The Swamped Common-Emitter Amplifier

OBJECTIVE

To observe the effect on voltage gain and stability of the swamped CE amplifier.

THEORY

The bypassed CE amplifier was analyzed in Electronics Exercise 5. The purpose of the emitter-bypass capacitor is to prevent degenerative feedback of the AC signal in the emitter resistor required for DC bias. A compromise of improved stability and higher input impedance against voltage gain can be achieved by partially bypassing the emitter resistor. In Figure 6.1, the two emitter resistors make up the total emitter resistance and only one is bypassed.

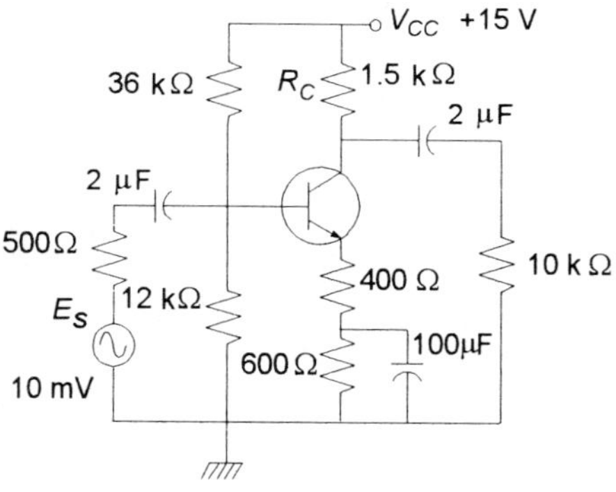

Figure 6.1

PROCEDURE

The circuit drawing of Figure 6.2 is the *Breadboard* equivalent of the swamped-emitter CE amplifier in Figure 6.1. Draw and solve the circuit and obtain the solution for a frequency of 1.0 kHz. Obtain a printout of the schematic and mark on the circuit the current through each branch.

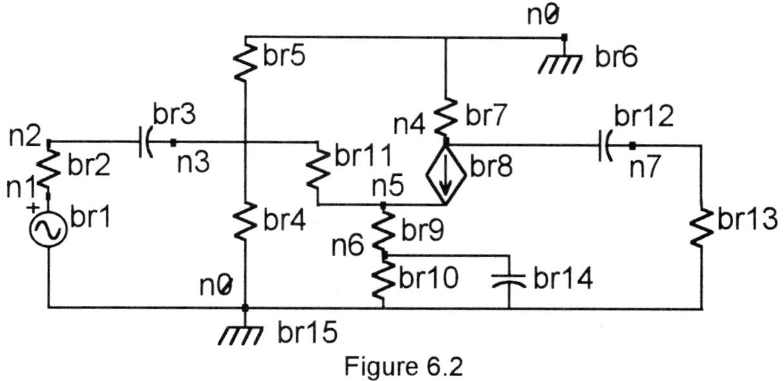

Figure 6.2

Net List for the Circuit

Br. 1: E = 10 mV/0°	Br. 7: R = 1.5 kΩ	Br. 11: R = 3.5 kΩ
Br. 2: R = 500 Ω	Br. 8: CCCS, beta = 100	Br. 12: C = 2.0 μF
Br. 3: C = 2.0 μF	control branch 11	Br. 13: R = 10 kΩ
Br. 4: R = 12 kΩ	Br. 9: R = 400 Ω	Br. 14: C = 100 μF
Br. 5: R = 36 kΩ	Br. 10: R = 600 Ω	Br. 15: voltage reference
Br. 6: voltage reference		

Calculate the voltage gain of the amplifier.

$$A_v = (V_{node\ 7})/(V_{node\ 2}) = \underline{\hspace{5cm}}$$

Compare to A_v from Electronics Exercise 5: $A_v = \underline{\hspace{4cm}}$

Has the voltage gain decreased? $\underline{\hspace{3cm}}$

Calculate the input impedance of the amplifier.

$$Z_{in} = (V_{node\ 2})/(I_{branch\ 2}) = \underline{\hspace{4cm}}\ \Omega$$

Compare to Z_{in} of Electronics Exercise 5: $Z_{in} = \underline{\hspace{3cm}}\ \Omega$

Calculate the current gain of the amplifier.

$$A_i = (I_{branch\ 13})/(I_{branch\ 2}) = \underline{\hspace{4cm}}$$

Calculate the output impedance by the method of Thevenin's theorem. (See page 146 for the procedure.)

$$Z_{out} = E_{oc}/I_{sc} = \underline{\hspace{3cm}}\ \Omega$$

Comment on the change of voltage gain and input impedance, comparing the results above to the common-emitter amplifier of Electronics Exercises 5.

Comment

FURTHER ANALYSIS

Suppose the transistor is replaced with one of the same type but having $h_{fe} = 200$ and $h_{ie} = 1.0$ kΩ (branch 11). Edit the circuit drawing for the new values and determine the voltage gain and input impedance. Comment on the stability of the amplifier by comparing the gain and input impedance of the two amplifiers.

Common-Collector (Emitter-Follower) Amplifier

OBJECTIVE

To analyze a common-collector amplifier and to determine the parameters of interest.

THEORY

The emitter-follower configuration of a transistor amplifier obtains its output signal from across the emitter resistor. The voltage gain of the amplifier is never larger than unity and the output signal is in phase with the input signal. A current gain is realized that gives the amplifier an overall power gain.

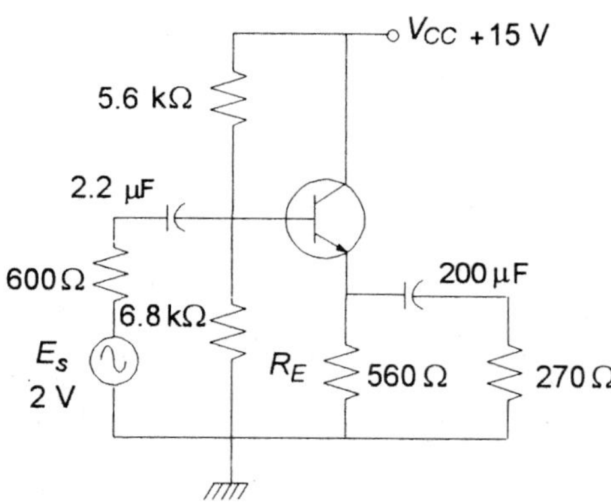

Figure 7.1

The advantage of the common-collector amplifier is that the input impedance is lower than that of the CE amplifier. This will provide for impedance matching to a low-impedance source thereby maximizing power transfer. In some situations a voltage gain is not needed but an overall power gain is required. The common-collector amplifier is optimum for this situation. The CC amplifier can be readily recognized since it has no collector resistor and the emitter resistor is not bypassed.

PROCEDURE

Figure 7.1 shows the schematic diagram for a common-collector amplifier and Figure 7.2 shows the equivalent circuit model.

Using *Breadboard* in the AC mode, draw the circuit of Figure 7.2 and solve for a frequency of 1 kHz.

Net List For the Circuit

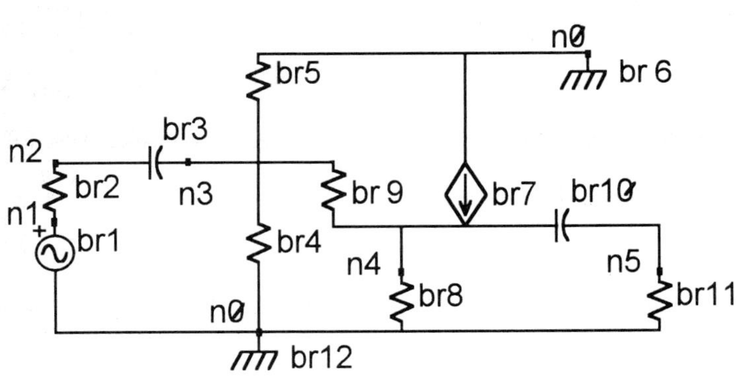

Figure 7.2

Br. 1: $E = 2$ V/0°
Br. 2: $R = 600\ \Omega$
Br. 3: $C = 2.2\ \mu F$
Br. 4: $R = 6.8\ k\Omega$
Br. 5: $R = 5.6\ k\Omega$
Br. 6: voltage reference
Br. 7: CCCS, beta = 100
         control branch 9
Br. 8: $R = 560\ \Omega$
Br. 9: $R = 2600\ \Omega$
Br. 10: $C = 200\ \mu F$
Br. 11: $R = 270\ \Omega$
Br. 12: voltage reference

Calculate the voltage gain of the amplifier.

$$A_v = (V_{out}/V_{in}) = (V_{node\ 5})/(V_{node\ 2}) = \underline{\hspace{2cm}}$$

Calculate the current gain of the amplifier.

$$A_i = I_{out}/I_{in} = (I_{branch\ 11})/(I_{branch\ 2}) = \underline{\hspace{2cm}}$$

Calculate the input impedance of the amplifier.

$$Z_{in} = (V_{node\ 2})/(I_{branch\ 2}) = \underline{\hspace{2cm}}\ \Omega$$

What are the voltage drops across the coupling capacitors?

$$V_{branch\ 3} = \underline{\hspace{1.5cm}}\ V \qquad\qquad V_{branch\ 10} = \underline{\hspace{1.5cm}}\ V$$

Follow the procedure on page 146 to determine the output impedance of the CC amplifier.

$$\text{open-output voltage} = E_{oc} = \underline{\hspace{2cm}}\ V$$

$$\text{short-circuit current } I_{sc} = \underline{\hspace{2cm}}\ mA$$

The output impedance of the CC amplifier $\qquad Z_{out} = E_{oc}/I_{sc} = \underline{\hspace{1.5cm}}\ \Omega$

What is the power gain of the amplifier?

$$A_p = P_o/P_{in} = A_v A_i = \underline{\hspace{2cm}}$$

Compare Z_{in} and Z_{out} to the values obtained for the CE amplifier.

FURTHER ANALYSIS

Compare the results obtained above to those obtained by the formulas in your text.

BJT Amplifier Frequency Response

OBJECTIVE

To obtain the frequency response curves of a BJT amplifier and to investigate the causes of the lower cutoff frequency.

THEORY

The low-frequency cutoff of a transistor amplifier is determined by the size of the coupling capacitors and the emitter-bypass capacitor. In combination with the resistors of the circuit, these capacitors form high-pass filter circuits, which attenuate the low-frequency signals. The lowest cutoff frequency is determined by the R-C circuit with the highest cutoff frequency.

At high frequencies the base-collector and base-emitter junction capacitances of the BJT, although small in value, become significant. The value of the C_{cb} capacitor is given on manufacturer's data sheets as C_{ob} or C_{obo}. The value of the C_{be} capacitor may be calculated from the relationship $C_{be} = 1/(2\pi f_T r'_e)$ where f_T is the current gain-bandwidth product for the transistor and $r'_e = h_{ie}/h_{fe}$. In addition, wiring (stray) capacitance due to placement of the connecting wires also affects the frequency response.

PROCEDURE

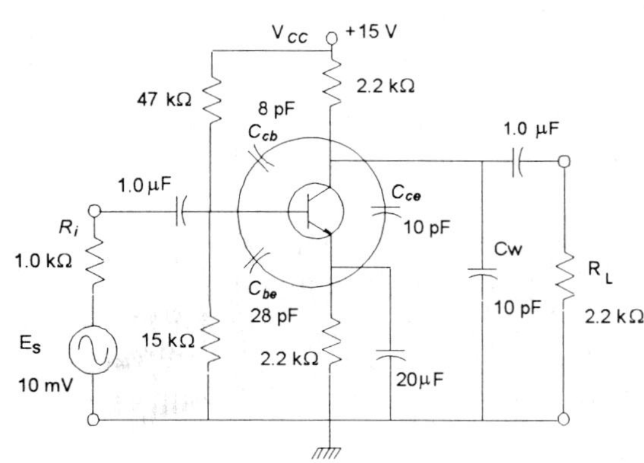

Figure 8.1

Figure 8.1 is the circuit diagram for a CE amplifier. Note the capacitors that have been added to simulate the C_{cb}, C_{be}, and C_{ce} junction capacitance and the estimated wiring capacitance (C_w) of 10 pF.

Figure 8.2 is the model of the CE amplifier of Figure 8.1. Using the AC analysis option of *Breadboard*, draw the circuit model for the values of components shown in the net list. Using the graph option, obtain a graph of the voltage output (node 6) for a frequency range of 100 Hz to 10 MHz.

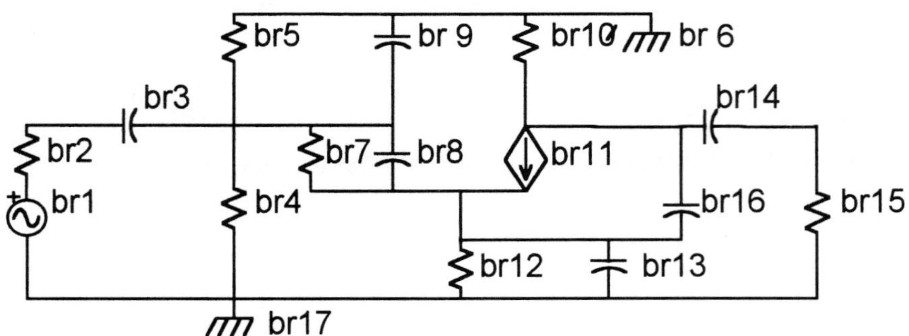

Figure 8.2
Net List for the Circuit

Br. 1: E = 10 mV   Br. 7: R = 1.9 kΩ   Br. 12: R = 2.2 kΩ
Br. 2: R = 1 kΩ   Br. 8: C = 28 pF   Br. 13: C = 20 μF
Br. 3: C = 1.0 μF   Br. 9: C = 8 pF   Br. 14: C = 1 μF
Br. 4: R = 15 kΩ   Br. 10: R = 5.6 kΩ   Br. 15: R = 2.2 kΩ
Br. 5: R = 47 kΩ   Br. 11: CCCS, beta = 100   Br. 16: C = 10 pF
Br. 6: voltage reference   control branch 7   Br. 17: voltage reference

From the graph, find the frequencies at which the output voltage has dropped to
0.707 the voltage at midrange. These are the half-power frequencies that define the
bandwidth BW $= (f_2 - f_1)$.

$f_1 =$ _______ Hz $f_2 =$ _______ Hz   BW $= (f_2 - f_1) =$ _________ Hz

What is the voltage gain at mid-frequency $(f_2 - f_1)/2$?

$A_v = (V_{out}/V_{in})$ ________________

Edit the *Breadboard* circuit drawing and change the emitter-bypass capacitor to
200 μF. Obtain the frequency response graph for the same range of frequencies.

$f_1 =$ _______ Hz $f_2 =$ _______ Hz   BW $= (f_2 - f_1) =$ _______ Hz

Edit the *Breadboard* circuit drawing and change the coupling capacitors to 10 μF.
Obtain the frequency response graph for the same range of frequencies.

$f_1 =$ _______ Hz $f_2 =$ _______ Hz   BW $= (f_2 - f_1) =$ _______ Hz

Comment on the effect of changing the values of the coupling capacitors and the
emitter bypass capacitor.

Comment

JFET DC Model and Self-Bias

OBJECTIVE

To model the field-effect transistor and to apply the model to study a JFET self-bias circuit.

THEORY

The approximate DC model of a JFET is shown in Figure 9.1.

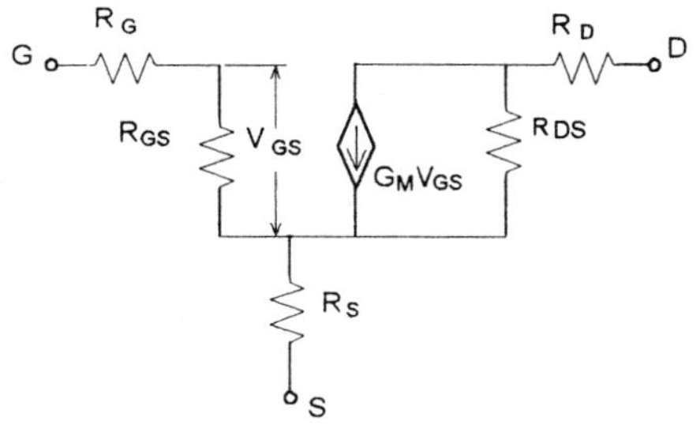

Figure 9.1

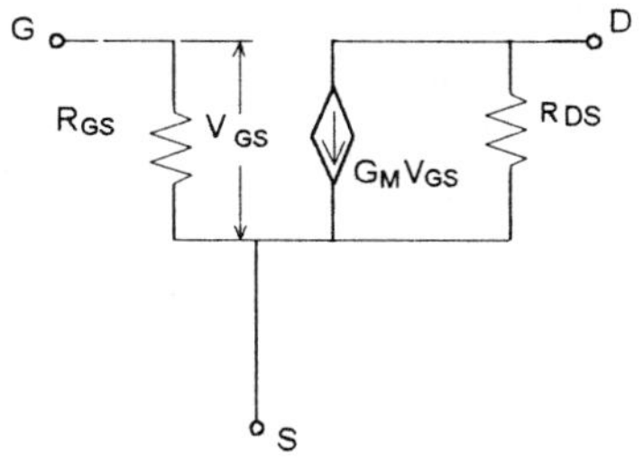

Figure 9.2

The electronic amplifying characteristics of the JFET are represented by the voltage-controlled current source (VCCS) which causes a current to flow whose magnitude is G_M (the transconductance of the FET) times as large as the gate-source voltage V_{GS}. The input resistance of the FET (R_{GS}) representing the reversed-biased gate-source diode is of the order of 10^9 Ω. The resistors R_S, R_G, and R_D represent the bulk resistance of the semiconductor material and are of the order of 1 Ω to 10 Ω; they can be eliminated from the model. Also the gate-drain resistance (R_{GD}) is of the order of 10^9 Ω and can be left out of the model. The drain-source output resistance R_{DS} is of the order of 100 kΩ to 1 MΩ and since it will be paralleled by a much smaller drain resistor (R_D), can be removed from the model. The simplified ideal model for the JFET is shown in Figure 9.2.

Breadboard's circuit-analysis capabilities are restricted to the linear operating region of the FET's I_D versus V_{DS} characteristic curves where the drain-source voltage is greater than the pinch-off voltage and less then the breakdown voltage ($V_P < V_{DS} < V_{BR}$).

Manufacturers' specification sheets provide values for the $V_{GS(off)}$, I_{DSS}, and g_{mo} of the device. In modeling an FET for DC bias, the DC transconductance ($G_M = I_D/V_{GS}$) of the device must be used. The G_M varies with the operating point (Q) on the transconductance curve and is given as:

$$G_M = (I_{DSS}/V_{GS})[\ 1 - (V_{GS}/V_{GS(off)})]^2$$

Self-bias is the easiest method of biasing an FET and simply involves connecting a large-value resistor from the gate to the chassis-common point. Since the input resistance of the FET is very large in value (100 MΩ or more), practically no current flows in the gate resistor, and the gate voltage V_G is held at 0 V. The source current flowing in the source resistor R_S causes the voltage at the source to be positive such that the gate-source voltage V_{GS} is negative. The magnitude of the gate-source voltage V_{GS} must be less than the pinch-off voltage of the FET in order for the Q-point to be in the operating range.

In modeling the self-bias JFET circuit, both the G_M and the V_{GS} must be specified. The transconductance curve of Figure 9.4 shows that as V_{GS} is reduced in magnitude the drain current increases and the G_M also increases. For midpoint bias, the G_M as calculated from the equation above is $G_M = I_{DSS}/(2V_{GS(off)})$.

If the FET is replaced by another of the same type that has a different I_{DSS} and $V_{GS(off)}$, the Q-point will shift but still remain within the operating region of the device due to the self-regulating feature of the circuit.

PROCEDURE

Figure 9.3 depicts a JFET self-bias circuit whose parameters are $G_M = 1000$ uS, $I_{DSS} = 16$ mA, and $V_{GS(off)} = -8$ V. Figure 9.4 is a graph of the minimum and maximum transconductance curves for the transistor and the load-line for the circuit.

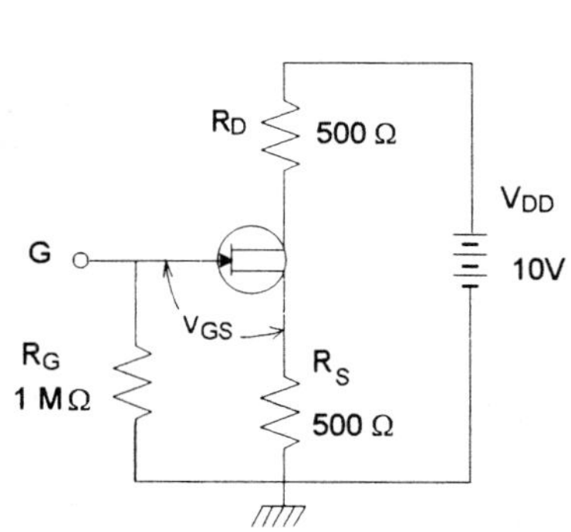

Figure 9.3

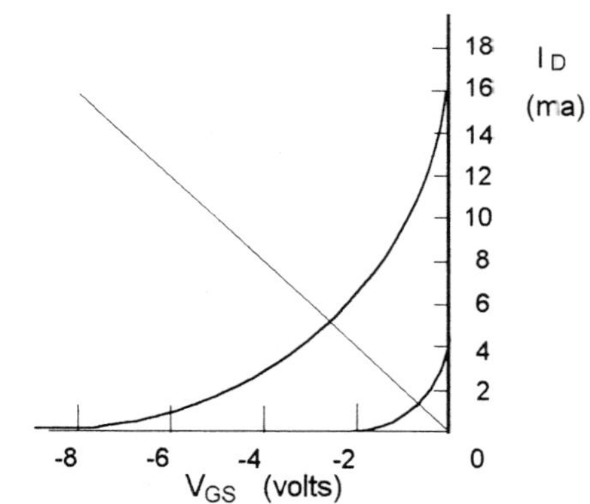

Figure 9.4

Using *Breadboard* in the DC analysis mode, draw the circuit model shown in Figure 9.5 for the values given in the net list. Note that the input resistance of the FET (100 MΩ) has been included in the model to provide the control-branch voltage for the VCCS. The control branch (branch 5) must be drawn with an upward cursor movement to provide the program with the correct current direction. The 4 V source (branch 7) is the V_{GS} for midpoint bias.

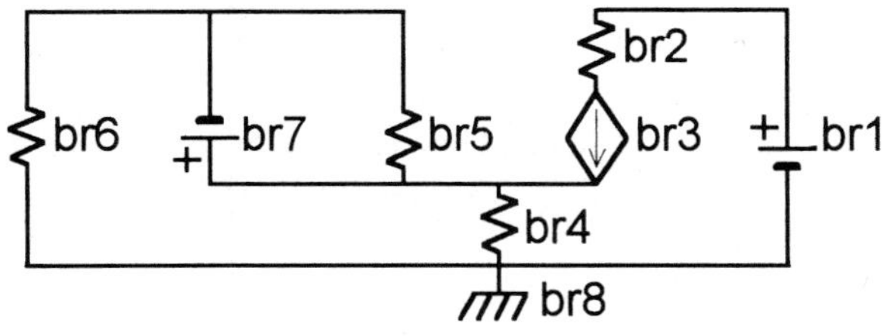

Net List for the Circuit

Br. 1: E = 20 V
Br. 2: R = 1 kΩ
Br. 3: VCCS, G_M = 2000 μS
      control branch 5
Br. 4: R = 1 kΩ
Br. 5: R = 100 MΩ
Br. 6: R = 1 MΩ
Br. 7: E = 4 V
Br. 8: voltage reference

Figure 9.5

Solve the circuit and from the results, mark on Figure 9.3 and Figure 9.5 the voltage and current for each branch of the circuit.

For G_M = 1000 μS, I_{DSS} = 16 mA, $V_{GS(off)}$ = –8 V, V_{GS} ($V_{branch\ 7}$) = –4 V:

$$V_G = V_{node\ 4} = \text{\_\_\_\_\_\_\_} \ \mu V \qquad V_{DS} = V_{branch\ 3} = \text{\_\_\_\_\_\_\_\_\_} \ V$$

$$I_D = I_{branch\ 3} = \text{\_\_\_\_\_\_\_\_} \ mA \qquad I_G = I_{branch\ 5} = \text{\_\_\_\_\_\_\_\_\_\_} \ nA$$

Power dissipated in the FET ($I_D V_{DS}$) P_D = __________ mW.

Total power dissipated by the circuit P_T = __________ mW.

On Figure 9.4, mark the operating point for the FET.

Suppose an FET with G_M = 1000 μS and $V_{GS(off)}$ = –6 V is used in the circuit. Edit the *Breadboard* drawing for the V_{GS} of the new FET and obtain the circuit solution.

For G_M = 1000 μS, I_{DSS} = 12 mA, $V_{GS(off)}$ = –6 V, V_{GS} ($V_{branch\ 7}$) = –3 V:

$$V_G = V_{node\ 4} = \text{\_\_\_\_\_\_\_} \ \mu V \qquad V_{DS} = V_{branch\ 3} = \text{\_\_\_\_\_\_\_\_\_} \ V$$

$$I_D = I_{branch\ 3} = \text{\_\_\_\_\_} \ mA \qquad I_G = I_{branch\ 5} = \text{\_\_\_\_\_\_\_\_\_\_} \ nA$$

On the load line of Figure 9.4, mark the operating point for the FET.

Repeat the above procedure for an FET with G_M = 1000 μS, I_{DSS} = 8 mA, $V_{GS(off)}$ = –4 V, and V_{GS} = –2 V.

For G_M = 1000 μS, I_{DSS} = 8 mA, $V_{GS(off)}$ = –4, V_{GS} ($V_{branch\ 7}$) = –2 V

$V_G = V_{node\ 4}$ = _______ μV $V_{DS} = V_{branch\ 3}$ = _________ V

$I_D = I_{branch\ 3}$ = _______ mA $I_G = I_{branch\ 5}$ = _________ nA

On Figure 9.4, mark the operating point for the FET.

Ideally, the drain current I_D should remain constant for each FET used. Comment on the stability of the operating point for FETs having different values of transconductance and $V_{GS(off)}$.

Comments

FURTHER ANALYSIS

Following your textbook on the procedure for designing a self-bias circuit for an FET, calculate the values of source and drain resistors using the formulas derived for self-bias. Confirm your design by using *Breadboard* to solve the model of the circuit.

JFET Voltage-Divider Bias

10

OBJECTIVE

To determine the operating points of a voltage-divider FET bias circuit for FETs of the same type but with different characteristics.

THEORY

To provide a more stable operating point than self-bias, voltage divider bias is commonly used. In this method the voltage at the gate of the FET is held constant and the voltage across the source resistor R_S is made large enough so that the wide tolerances of the FET parameters will not have a significant effect. The value of R_S will be in the range of kΩ to provide the necessary V_{GS}, and, since the gate draws essentially no current, the voltage-divider resistors are chosen to be large in value to maintain the high-input-impedance characteristic of FETs.

PROCEDURE

The circuit diagram of Figure 10.1 shows a voltage divider bias circuit for an FET. With voltage-divider bias, V_{GS} is established as the voltage difference between V_G and V_S. The G_M for the FET is calculated as $G_M = I_{DSS}/(2V_{GS(off)})$ for midpoint bias. Note that the input resistance of the FET (100 MΩ) has been included in the model to provide the control branch for the VCCS. The transconductance curves of Figure 10.2 were obtained from the maximum and minimum values of $V_{GS(off)} = -5$ V, $I_{DSS} = 10$ mA, $V_{GS(off)} = -2$ V, and $I_{DSS} = 4$ mA for the FET.

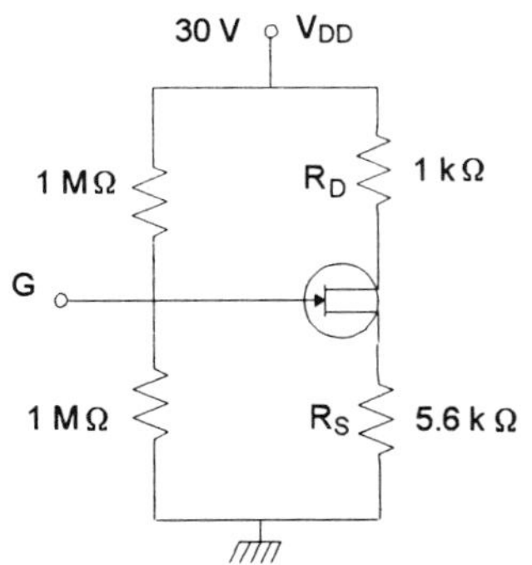

Figure 10.1

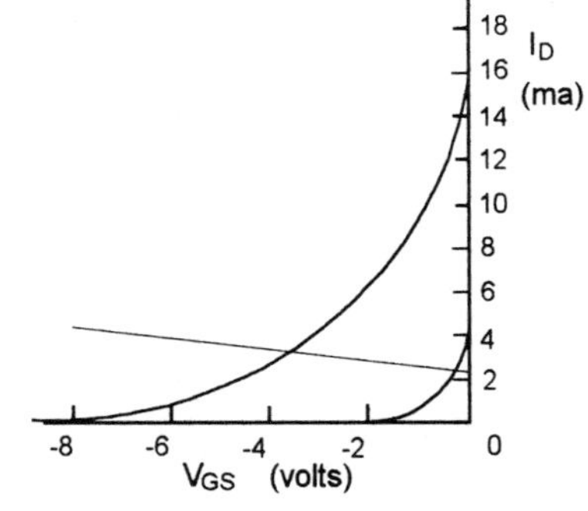

Figure 10.2

Using *Breadboard* in the DC analysis mode, draw the circuit model shown in Figure 10.3 for the values given in the net list. Note that branch 7, the control branch, must be drawn with an upward cursor movement to provide the program with the current direction in the resistor.

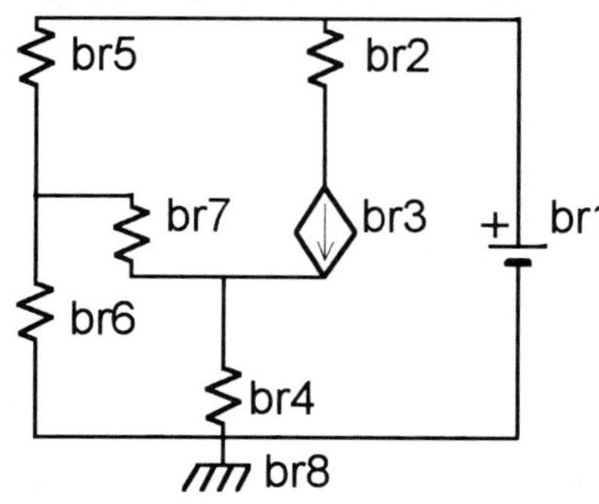

Figure 10.3

Net List for the Circuit

Br. 1: E = 30 V
Br. 2: R = 1.0 Ω
Br. 3: VCCS, G_M = 800 μS
    control branch 7
Br. 4: R = 5.6 kΩ
Br. 5: R = 1.0 MΩ
Br. 6: R = 1.0 MΩ
Br. 7: R = 100 MΩ
Br. 8: voltage reference

Solve the circuit and from the results mark on Figures 10.1 and 10.3 the voltage and current for each branch of the circuit.

$$V_{DS} = V_{branch\ 3} = \underline{\hspace{2cm}} \text{ V} \qquad V_{GS} = V_{branch\ 7} = \underline{\hspace{2cm}} \text{ V}$$

$$I_D = I_{branch\ 2} = \underline{\hspace{2cm}} \text{ mA} \qquad I_G = I_{branch\ 7} = \underline{\hspace{2cm}} \text{ }\mu\text{A}$$

Power dissipated by the FET ($I_D V_{DS}$) __________ mW

On the load line of Figure 10.2, mark the operating point for the FET.

Suppose an FET with a G_M of 1000 μS, (I_{DSS} = 12 mA, and $V_{GS(off)}$ = –6 V) is used in the circuit. Edit the *Breadboard* drawing for the new FET and obtain the circuit solution. Mark the new operating point on Figure 10.2.

Repeat the above for an FET with a G_M of 1400 μS, (I_{DSS} = 8 mA, and $V_{GS(off)}$ = –4 V).

Repeat the above for an FET with a G_M of 2700 μS, (I_{DSS} = 4 mA, and $V_{GS(off)}$ = –2 V).

Comment on the stability of the operating point for FETs having different values of transconductance.

Comments

Common-Source FET-Amplifier

OBJECTIVE

To model the FET amplifier and study the characteristics of an FET amplifier in the common-source configuration.

THEORY

The model of a JFET in the AC mode of operation is shown in Figure 11.1. The capacitors C_{gs}, C_{gd}, and C_{ds} represent the junction capacitance. These capacitances are in the order of a few picofarads and can be eliminated from the circuit model if the frequency of operation under study is below that at which they become significant (< 10 MHz). The resistor r_{gs} is the input resistance of the FET, which can be calculated from the manufacturer's specification sheet value of gate-reverse current at the specified V_{GS} voltage (for example, 15 V/1 nA = $15(10^9)$ Ω). The VCCS amplifies the voltage across r_{gs} to provide a current of $g_m v_{gs}$, where g_m is the transconductance of the FET and is defined as

$$g_m = \delta V_{GS}/\delta I_D \text{ Siemens.}$$

The transconductance of a FET is not constant and varies with the operating point determined by the bias circuit. The value of g_m can be calculated from the value of g_{mo} provided as a specification.

$$g_m = g_{mo}[1-V_{GS}/V_{GS(off)}]^{\frac{1}{2}} = g_{mo}(I_D/I_{DSS})^{\frac{1}{2}}$$

The output impedance of the FET (r_{ds}) can also be obtained as $1/y_{os}$ from the specifications. Above the pinch-off region the value of r_{ds} is in the megohm range, and since it is usually paralleled by a much smaller drain resistor (R_D), it may be eliminated from the model. The ideal model is shown in Figure 11.2.

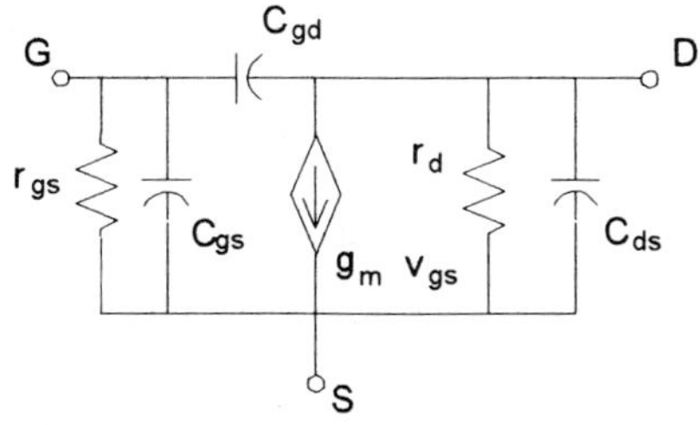

Figure 11.1

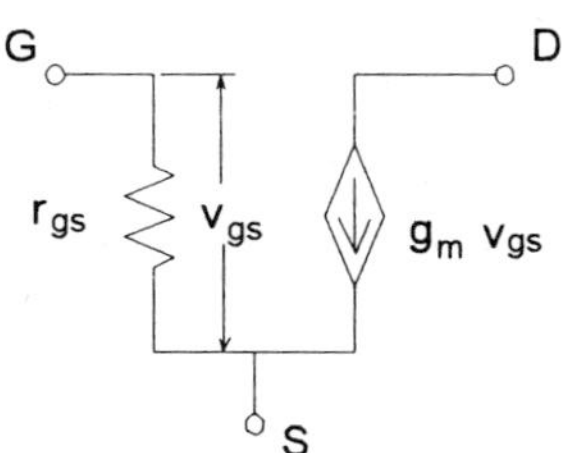

Figure 11.2

The common-source amplifier provides a high-input impedance and a low-noise figure. The common-source amplifier is often used in the front end of amplifiers in conjunction with an automatic gain control circuit.

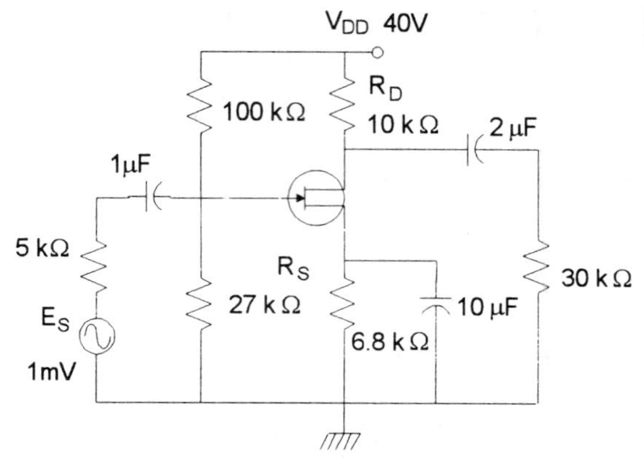

Figure 11.3

PROCEDURE

The common-source FET amplifier to be studied is shown in Figure 11.3, and its model is shown in Figure 11.4. Using *Breadboard* in the AC mode, draw the circuit model and solve for a frequency of 50 kHz.

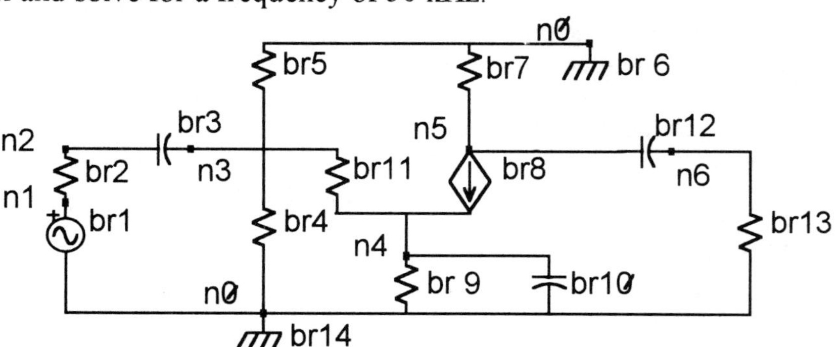

Figure 11.4

Net List for the Circuit

Br. 1: E = 1.0 mV
Br. 2: R = 5 kΩ
Br. 3: C = 1.0 µF
Br. 4: R = 27 kΩ
Br. 5: R = 100 kΩ
Br. 6: voltage reference
Br. 7: R = 10 kΩ

Br. 8: VCCS, g_m = 2000 µS
    control branch 11
Br. 9: R = 6.8 kΩ
Br. 10: C = 10 µF
Br. 11: R = 100 MΩ
Br. 12: C = 2.0 µF
Br. 13: R = 50 kΩ
Br. 14: voltage reference

Determine the voltage gain of the amplifier $(V_{node\ 6}/V_{node\ 1})$.

$A_v = (V_{node\ 6})/(V_{node\ 1}) = $ ___________________

By formula, $A_v = g_m(R_D\|R_L) = $ ___________________

Determine the current gain $A_i = I_{out}/I_{in}$.

$$A_i = (I_{branch\ 13})/(I_{branch\ 1}) = \underline{\hspace{3cm}}$$

Determine the input impedance of the amplifier.

$$Z_{in} = V_{in}/I_{in} = \underline{\hspace{3cm}} k\Omega$$

Determine the output impedance by the following method. Replace the load resistor, branch 13, with a very high value resistor representing an open circuit. Solve to obtain E_{oc}, the Thevenin voltage. Replace the output resistor with a very low value resistor representing a short circuit. Solve to obtain the current through the resistor.

$$Z_{out} = E_{oc}/I_{sc} \underline{\hspace{3cm}} k\Omega$$

Refer to your text for the formulas to calculate Z_{in} and Z_{out}. Perform the calculations for the values in this circuit and compare the results obtained by formula to those obtained above.

Calculations

FURTHER ANALYSIS

By leaving all or part of the source resistor (R_S) un-bypassed by the source capacitor (branch 10), the voltage gain can be made more predictable for FETs having differing g_{mo}, but the voltage gain will be reduced as a trade-off due to the degenerative feedback of the voltage across the un-bypassed portion of the source resistor.

Repeat the above analysis for the partially bypassed (swamped) source resistor configuration and comment on the voltage gain for the two cases.

Comment

Common-Drain FET-Amplifier

12

OBJECTIVE

To study the characteristics of a common-drain amplifier with regard to the voltage and current gain and the input and output impedance.

THEORY

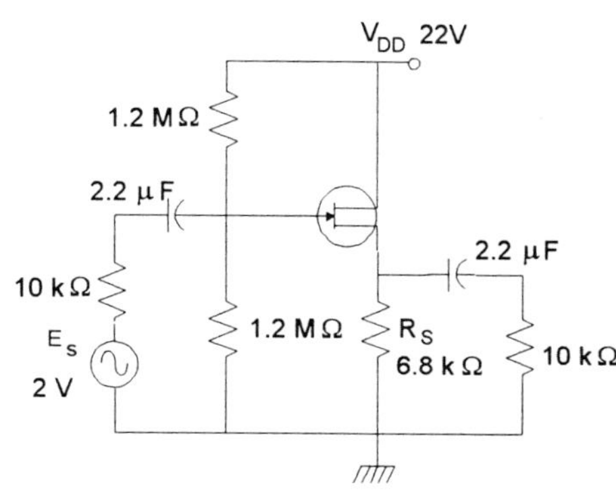

Figure 12.1

The output signal of the common-drain (source-follower) amplifier is taken across the source resistor as shown in Figure 12.1 and is in phase with the source voltage. The voltage gain of the amplifier is less than unity. This amplifier configuration is used as an impedance-matching buffer, since it provides very high input impedance and low output impedance.

Net List for the Circuit

Br. 1: E = 2.0 V
Br. 2: R = 10 kΩ
Br. 3: C = 2.2 μF
Br. 4: R = 1.2 MΩ
Br. 5: R = 1.2 MΩ
Br. 6: voltage reference
Br. 7: VCCS, g_m = 2500 μS
    control branch 9
Br. 8: R = 6.8 kΩ
Br. 9: R = 100 MΩ
Br. 10: C = 2.2 μF
Br. 11: R = 10 kΩ
Br. 12: voltage reference

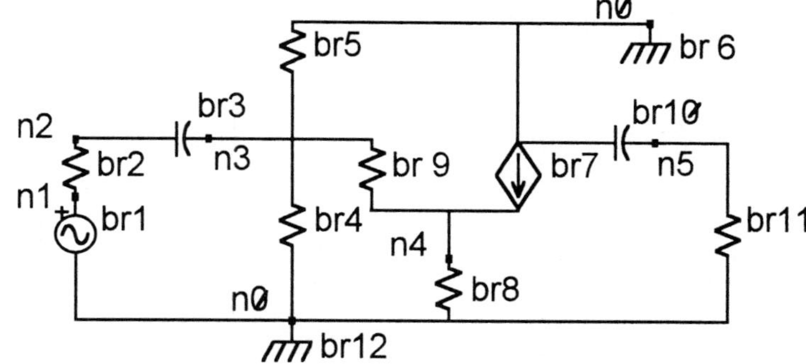

Figure 12.2

PROCEDURE

Using *Breadboard* in the AC mode, draw the circuit model shown in Figure 12.2 for the component values given in the net list. Solve the circuit for a frequency of 10 kHz and mark on the circuit diagram the current through each component and the voltage at each node.

The voltage gain of the amplifier is:

$$A_v = (V_{node\ 5})/(V_{node\ 1})\ \underline{\hspace{3cm}}$$

By formula, the voltage gain is:

$$A_v = r_s/(r_s + 1/g_m) = \underline{\hspace{3cm}} \quad \text{where } r_s = R_S \| R_L$$

The current gain of the amplifier is:

$$A_i = (I_{branch\ 11})/(I_{branch\ 1}) = \underline{\hspace{3cm}}$$

The input impedance of the amplifier is:

$$Z_{in} = (V_{node\ 1})/(I_{branch\ 1}) = \underline{\hspace{3cm}} \ \Omega$$

$$\text{By formula, } Z_{in} = R_S + R_1 \| R_2 = \underline{\hspace{3cm}} \ \Omega$$

Calculate the output impedance of the amplifier by Thevenin's theorem. Replace the load resistor, branch 11, with a very high value resistor representing an open circuit. Solve to obtain E_{oc}, the Thevenin voltage. Replace the output resistor with a very low value resistor representing a short circuit. Solve to obtain the current through the resistor.

$$Z_{out} = E_{oc}/I_{sc} = \underline{\hspace{3cm}} \ \Omega$$

$$\text{By formula, } Z_{out} = R_S \| (1/g_m) = \underline{\hspace{3cm}} \ \Omega$$

FURTHER ANALYSIS

The value of the load resistor will affect the voltage gain of the amplifier. Change the load resistor to 1 kΩ and repeat the above analysis.

High-Frequency FET-Amplifier Response

13

OBJECTIVE

To determine the effect of the junction capacitances on the voltage gain of the FET amplifier.

THEORY

At high frequencies, the junction capacitances of the FET and the capacitance of the connecting wires of the circuit, although small in value (pF), reduce the voltage output of the circuit. These capacitances limit the maximum frequency at which the amplifier can be used. The circuit in Figure 13.1 shows the shunting effect of these capacitances diagrammatically. The wiring capacitance is an estimated value.

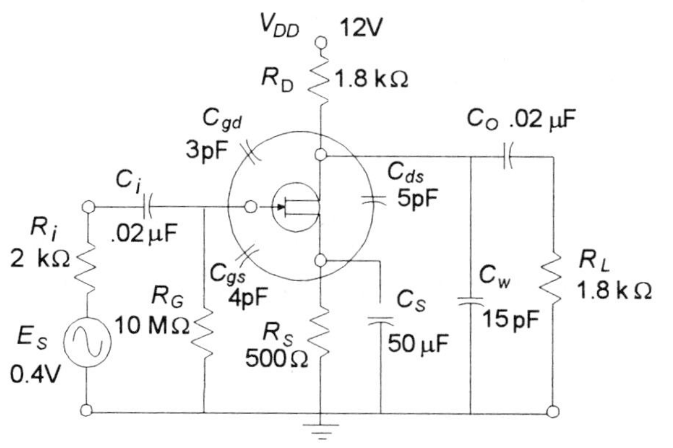

Figure 13.1

PROCEDURE

The model for this circuit is shown in Figure 13.2. The capacitances of branch 3 and branch 11 are the coupling capacitors. The capacitors of branches 5, 10, and 14 represent the C_{gd}, C_{gs}, and C_{ds} junction capacitances, respectively. The capacitance of branch 15 is the estimated wiring capacitance.

The voltage-controlled current source (VCCS) representing the FET and its power source has its current controlled by the voltage across branch 4. The g_m of the FET is given as 3.3 mS, and its input resistance is 10 MΩ. The output resistance r_o is 40 kΩ and has not been included, because it would be in parallel with the much smaller load resistance of 1.8 kΩ.

Net List for the Circuit

Br. 1: E = 0.4 V/0°   Br. 9: R = 500 Ω
Br. 2: R = 2 kΩ   Br. 10: C = 3 pF
Br. 3: C = 0.02 μF   Br. 11: C = 0.2 μF
Br. 4: R = 10 MΩ   Br. 12: R = 1.8 kΩ
Br. 5: C = 4 pF   Br. 13: C = 50 μF
Br. 6: voltage reference   Br. 14: C = 5 pF
Br. 7: R = 1.8 kΩ   Br. 15: C = 15 pF
Br. 8: VCCS, g_m = 3300 μS   Br. 16: voltage reference
    control branch 4

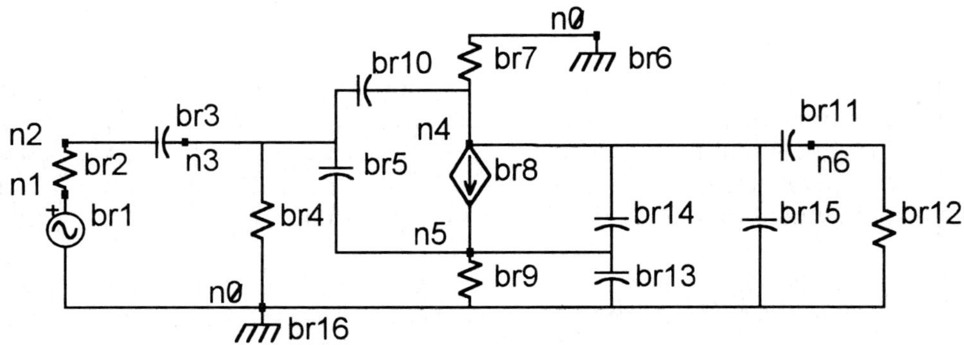

Figure 13.2

Using *Breadboard* draw and solve the circuit at a frequency of 100 kHz.

The voltage output (node 6) is __________ V at ________ °.

Solve the circuit at a frequency of 5 MHz.

The voltage output (node 6) is __________ V at ________ °.

For the change in frequency from 100 kHz to 5 MHz, the output voltage magnitude has dropped from _________ V to _________ V.

FURTHER ANALYSIS

Obtain a graph of the voltage output (node 6) as the frequency is varied over the frequency range of 10 kHz to 10 MHz. Determine the high-frequency cutoff.

The cutoff frequency of the FET amplifier circuit is ____________ Hz .

The phase shift at the cutoff frequency is __________ °.

Modeling the DC Operational Amplifier

OBJECTIVE

To model the operational amplifier (DC) and to confirm the closed-loop gain, input resistance, and output resistance.

THEORY

Op-amps are integrated circuits that allow the user to establish the gain and other characteristics for a wide variety of applications. The operational amplifier is nearly the ideal amplifier because it provides high input resistance, a very high voltage gain, and a low output resistance. With reference to the 741 op-amp, the input resistance is typically 2 MΩ, the open-loop gain is approximately 200,000, and the output resistance approximately 75 Ω. Op-amps are generally used with negative feedback to establish the desired voltage gain and to obtain stability of the voltage gain, which would otherwise be affected by temperature variation and changes in supply voltage. Op-amps may be connected in several different configurations. In this exercise the non inverting, the inverting, and the voltage follower circuits are analyzed.

PROCEDURE Part 1 – Non-inverting Amplifier

Figure 14.1 is the schematic circuit for a non-inverting operational amplifier circuit. The model for the circuit in the DC mode is shown in Figure 14.2, where the amplifying capability of the op-amp is provided by the VCVS. The voltage of the VCVS is dependent on the magnitude of the voltage across the input resistance (R_i). The negative voltage feedback is determined by the voltage divider circuit made up of R_f (branch 7) and R_1 (branch 3). The polarity of the voltage output of the VCVS is established by choosing the appropriate VCVS polarity symbol when drawing the circuit using *Breadboard*.

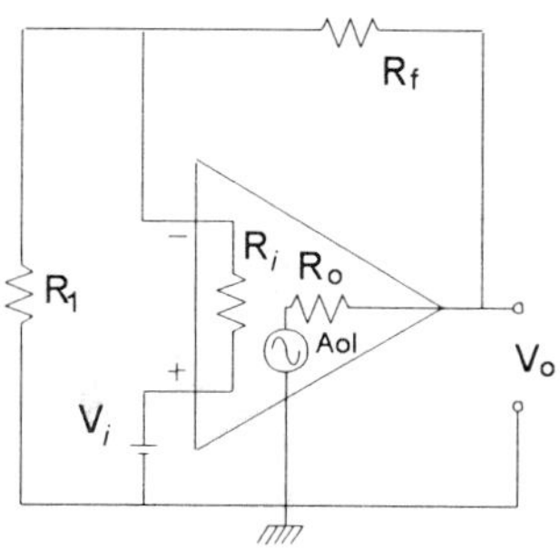

Figure 14.1

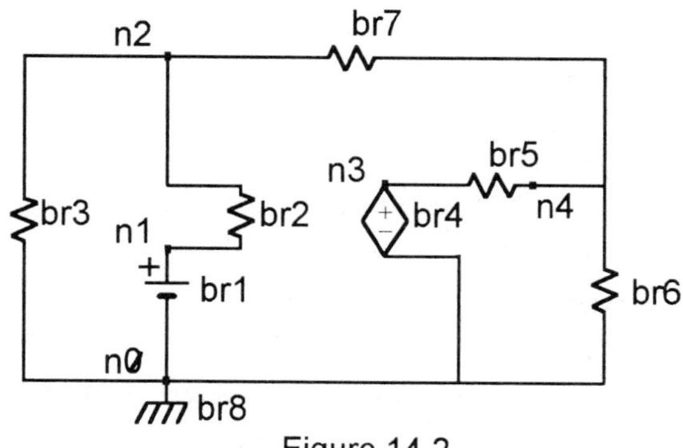

Figure 14.2

Draw the circuit model shown in Figure 14.2 using *Breadboard* in the DC mode for the values given in the net list for the circuit and solve the circuit.

Net List for the Circuit

Br. 1: E = 5 mV

Br. 2: R = 2 MΩ

Br. 3: R = 1 kΩ

Br. 4: VCVS, A_v = 200,000 control branch 2

Br. 5: R = 75 Ω

Br. 6: R = 10 kΩ

Br. 7: R = 100 kΩ

Br. 8: voltage reference

Mark on the circuit diagram all the branch currents and node voltages.

In developing the formulas for gain and input and output resistance of the op-amp non inverting amplifier, two assumptions are made.

The voltage across the input of the op-amp is approximately zero.

$$V_{branch\ 2} = \underline{\hspace{4cm}} V$$

The input current is approximately zero.

$$I_{in} = I_{branch\ 2} = \underline{\hspace{4cm}} A$$

Are the assumptions reasonable? $\underline{\hspace{3cm}}$

From the solution results, the closed loop voltage gain of the circuit is:

$$A_{cl} = V_{out}/V_{in} = (V_{node\ 4})/(V_{node\ 1}) = \underline{\hspace{4cm}}$$

By formula, $A_{cl} = 1 + R_f/R_1 = \underline{\hspace{4cm}}$

From the solution results, the input resistance R'_{in} can be calculated as:

$$E/(I_{branch\ 2}) = \underline{\hspace{4cm}} \Omega$$

By formula, $R'_{in} = R_i(1 + \alpha A_{ol}) = \underline{\hspace{4cm}} \Omega$

$$[\alpha = R_1/(R_1 + R_f)]$$

From the solution results, the output resistance R'_{out} can be obtained as: (See DC Exercise 13 for the procedure for finding E_{oc} and I_{sc})

$$E_{oc}/I_{sc} = \underline{\hspace{3cm}} \Omega$$

By formula, $R'_{out} = R_o/(1 + \alpha A_{ol}) = \underline{\hspace{3cm}} \Omega$

$$[R_o = 75\ \Omega]$$

PROCEDURE Part 2 – Inverting Amplifier

The circuit for an inverting amplifier is shown in Figure 14.3 and its model is shown in Figure 14.4. Note the polarity of the VCVS to simulate inversion of the input signal by the amplifier.

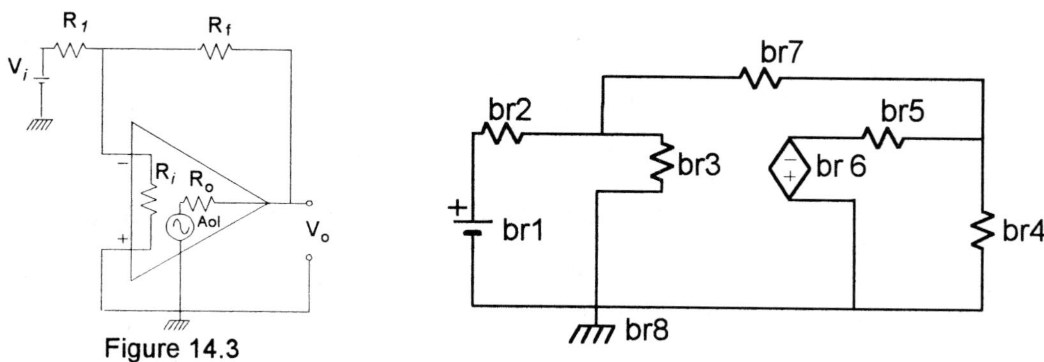

Figure 14.3

Figure 14.4

Draw and solve the circuit using *Breadboard* in the DC mode for the values given in the net list. On the circuit diagram of Figure 14.4, mark the node voltages and the currents in each branch.

Net List for the Circuit

Br. 1: E = 5 mV
Br. 2: R = 1 kΩ
Br. 3: R = 2 MΩ
Br. 4: R = 10 kΩ
Br. 5: R = 75 Ω

Br. 6: VCVS, A_v=200,000
     control branch 3
Br. 7: R = 100 kΩ
Br. 8: voltage reference

The voltage across the input of the op-amp is approximately zero. $V_{branch\ 3}$ =____V.

The current into the op-amp is approximately zero. $I_{branch\ 3}$ = ________ A

From the solution results, the closed-loop voltage gain of the circuit is:

$$A_{cl} = V_{out}/V_{in} = (V_{node\ 4})/(V_{node\ 1}) = \underline{\hspace{3cm}}$$
$$\text{By formula, } A_{cl} = R_f/R_1 = \underline{\hspace{3cm}}$$

From the solution results, the input resistance R'_{in} can be calculated as

$$(V_{branch\ 1})/(I_{branch\ 1}) = \underline{\hspace{2cm}} \ \Omega$$
$$\text{By formula, } R'_{in} = R_1 = \underline{\hspace{2cm}} \ \Omega$$

From the solution results, the output resistance R'_{out} can be obtained as:

$$E_{oc}/I_{sc} = \underline{\hspace{3cm}} \ \Omega$$
$$\text{By formula, } R'_{out} = R_o/(\alpha A_{ol}) = \underline{\hspace{2cm}} \ \Omega \text{ where } (R_o = 75 \ \Omega)$$

PROCEDURE Part 3 – Voltage Follower

The voltage follower configuration provides a very high input resistance and a very low output resistance, but the voltage gain is approximately unity. It is used for special circuits requiring a buffer such as an electronic voltmeter or electronic thermometer. A direct connection is made from the output to the inverting input of the op-amp, as shown in Figure 14.5. With *Breadboard,* a very low value resistor (0.0001 Ω) is used for the direct connection to provide a component for the calculation of current.

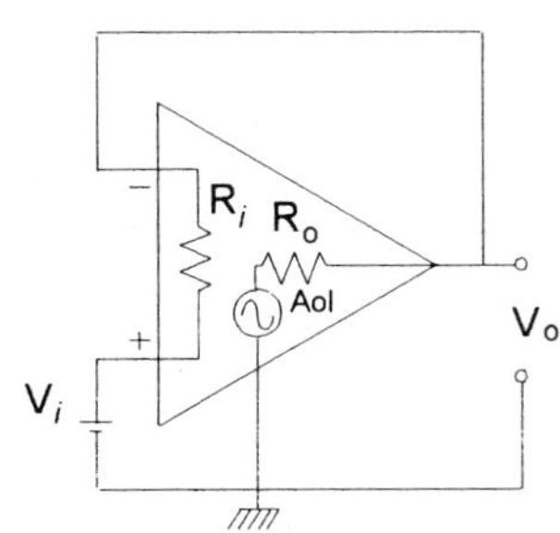

Figure 14.5

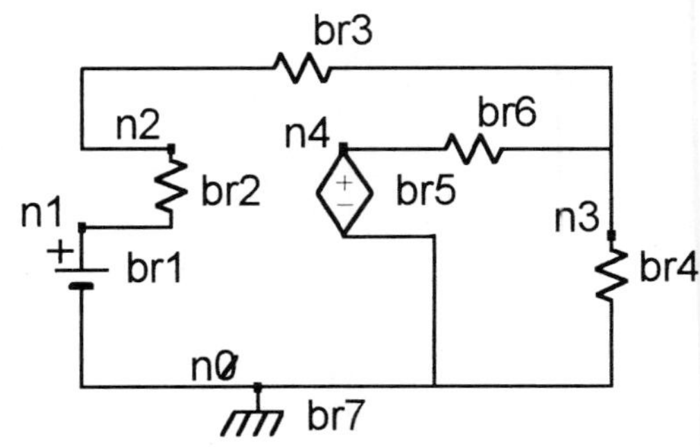

Figure 14.6

Net List for the Circuit

Br. 1: E = 1000 mV   Br. 5: VCVS, A_v = 200,000
Br. 2: R = 2 MΩ   control branch 2
Br. 3: R = 100 µΩ   Br. 6: R = 75 Ω
Br. 4: R = 10 kΩ   Br. 7: voltage reference

On Figure 14.6, mark all the node voltages and the branch currents.

The input voltage of the op-amp is approximately zero. $V_{branch\ 2}$ = ________ V

The current into the op-amp is approximately zero. $I_{branch\ 2}$ = ____________ A

From the solution results, the closed-loop voltage gain of the circuit is:

$$A_{cl} = V_{out}/V_{in} = (V_{node\ 3})/(V_{node\ 1}) = \underline{\hspace{3cm}}$$

From the solution results, the input resistance R'_{in} can be calculated as:

$$(V_{branch\ 2})/(I_{branch\ 2}) = \underline{\hspace{3cm}}\ \Omega$$

By formula, $R'_{in} = R_{in} A_{ol} = \underline{\hspace{4cm}}\ \Omega$

The output resistance R'_{out} can be obtained as:

$$E_{oc}/I_{sc} = \underline{\hspace{3cm}}\ \Omega$$

By formula, $R'_{out} = R_o/A_{ol} = \underline{\hspace{4cm}}\ \Omega$

OP-AMP Roll-Off Characteristics

OBJECTIVE

To model the op-amp for AC operation and to study the gain-frequency and phase-frequency characteristics.

THEORY

Operational amplifiers operate over a frequency range from zero Hz (DC) to MHz. They are frequency-compensated either externally or internally to provide an open-loop gain characteristic that decreases (roll-off) by –20 dB per decade above the cutoff frequency. Since op-amps are most commonly used with negative feedback, the purpose of the frequency compensation is to maintain stability by preventing oscillations that might otherwise occur due to phase shifts of the output voltage of 180° or more .

In modeling an AC op-amp circuit, the user must be aware of the gain-bandwidth and phase shift characteristics of the op-amp as given in the manufacturers' specifications. For the gain of the circuit to be wholly a function of the feedback loop, the open-loop gain must be larger than the closed-loop gain at the frequency of operation. See Figure 15.1 for typical roll-off characteristic curves.

PROCEDURE

The op-amp circuit model in Figure 15.2 shows the 30 pF feedback-compensating capacitor as branch 8. The input capacitance of the op-amp has been neglected since its value is typically in the order of few pF. The value of the compensating capacitor will vary with the type of op-amp used and has been chosen as 30 pF for this exercise. To investigate the roll-off feature, draw the model in Figure 15.2 using *Breadboard* in the AC mode for the values given in the net list .

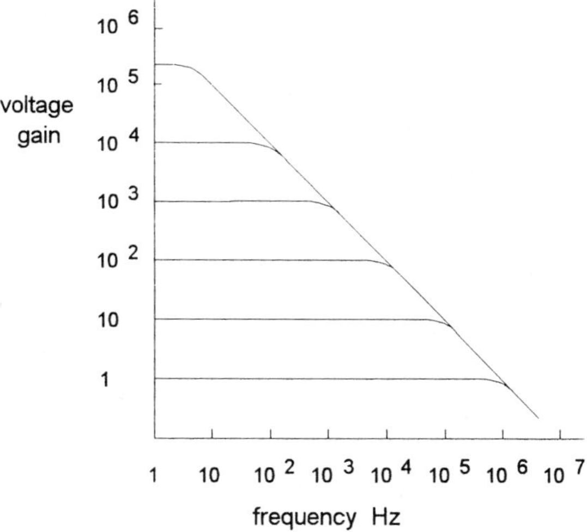

Figure 15.1

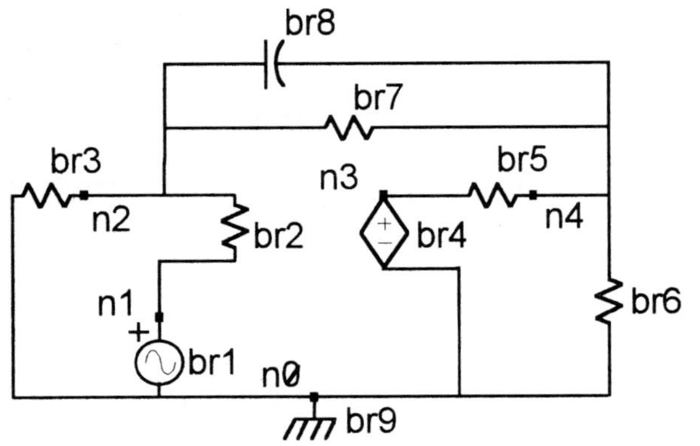

Figure 15.2

Net List for the Circuit

Br. 1: $E = 10$ mV/$0°$
Br. 2: $R = 2$ MΩ
Br. 3: $R = 1$ kΩ
Br. 4: VCVS, $A_v = 200{,}000$
      control branch 2
Br. 5: $R = 75$ Ω
Br. 6: $R = 10$ kΩ
Br. 7: $R = 1$ kΩ
Br. 8: $C = 30$ pF
Br. 9: voltage reference

Obtain a graph of the voltage output (node 4) over a frequency range of 10 kHz to 10 MHz using *Breadboard*.

From the graph, obtain the cutoff frequency. (The cutoff frequency occurs at the point where the output voltage has dropped to 0.707 of the maximum output.)

    cutoff frequency = ______________ Hz

Solve the circuit for the cutoff frequency and calculate A_V and $A_V f_C$.

$$A_V = (V_{node\ 4})/(V_{node\ 1}) = \text{\_\_\_\_\_\_\_\_} \qquad A_V f_C = \text{\_\_\_\_\_\_\_\_\_\_\_\_\_\_}$$

Solve the circuit for the frequencies given below and calculate the voltage gain for each of the frequencies and plot the values on Figure 15.1.

$$A_V = (V_{node\ 4})/(V_{node\ 1})$$

for 100 Hz $A_V = $ ___________________

for 10 kHz $A_V = $ ___________________

for 1 MHz $A_V = $ ___________________

Edit the circuit and change the feedback resistor (branch 7) to 10 kΩ.

Obtain a graph of the voltage output (node 4) over a frequency range of 1 kHz to 1 MHz using *Breadboard*.

From the graph obtain the cutoff frequency.

    cutoff frequency = ______________ Hz

Solve the circuit for the cutoff frequency and calculate A_V and $A_V f_C$.

$$A_V = (V_{node\ 4})/(V_{node\ 1}) = \text{\_\_\_\_\_\_\_\_} \qquad A_V f_C = \text{\_\_\_\_\_\_\_\_\_\_\_\_\_\_}$$

Solve the circuit for the frequencies given below, calculate the voltage gain for each of the frequencies, and plot the values on Figure 15.1. $A_V = (V_{node\ 4})/(V_{node\ 1})$

for 100 kHz $A_V =$ _______________________

for 1 MHz $A_V =$ _______________________

for 10 MHz $A_V =$ _______________________

Edit the circuit and change the feedback resistor to 100 kΩ.

Obtain a graph of the frequency response over the range of frequency from 100 Hz to 100 kHz.

cutoff frequency = _______________ Hz

Solve the circuit for the cutoff frequency and calculate A_V and $A_V f_C$.

$A_V = (V_{node\ 4})/(V_{node\ 1}) =$ _________ $A_V f_C =$ _______________

Solve the circuit for the frequencies given below and calculate the voltage gain for each of the frequencies and plot the values on Figure 15.1. $A_V = (V_{node\ 4})/(V_{node\ 1})$

for 10 Hz $A_V =$ _______________________

for 1 kHz $A_V =$ _______________________

for 100 kHz $A_V =$ _______________________

Edit the circuit and change the feedback resistor to 1 MΩ.

Obtain a graph of the frequency response over the range of frequency from 100 Hz to 100 kHz.

cutoff frequency = _______________ Hz

Solve the circuit for the cutoff frequency and calculate A_V and $A_V f_C$.

$A_V = (V_{node\ 4})/(V_{node\ 1}) =$ _________ $A_V f_C =$ _______________

Solve the circuit for the frequencies given below and calculate the voltage gain for each of the frequencies and plot the values on Figure 15.1. $A_V = (V_{node\ 4})/(V_{node\ 1})$

for 10 Hz $A_V =$ _______________________

for 100 Hz $A_V =$ _______________________

for 1 kHz $A_V =$ _______________________

Edit the circuit and change the feedback resistor to 10 MΩ.

Obtain a graph of the frequency response over the range of frequency from 1 Hz to 1 kHz.

cutoff frequency = _______________ Hz

Solve the circuit for the cutoff frequency and calculate A_V and $A_V f_C$.

$A_V = (V_{node\ 4})/(V_{node\ 1}) = $ _______ $A_V f_C = $ _______________

Solve the circuit for the frequencies given below and calculate the voltage gain for each of the frequencies and plot the values on Figure 15.1. $A_V = (V_{node\ 4})/(V_{node\ 1})$

for 1 Hz $A_V = $ _______________

for 10 Hz $A_V = $ _______________

for 100 Hz $A_V = $ _______________

In your own words, discuss the effect of roll-off with regard to frequency and gain of the operational amplifier. Why is the gain roll-off required in an op-amp circuit employing negative feedback?

Discussion

FURTHER ANALYSIS

How does the phase angle of the output voltage vary as a function of frequency? Using *Breadboard* to analyze the compensated op-amp circuit of Figure 15.2 for a closed-loop voltage gain of 100, record the phase angles of the output voltage. Write a description of the variation of the phase angle as a function of frequency.

Basic Active Filters

OBJECTIVE

To model a simple active low-pass, high-pass, or band-pass filter, and to obtain the response curve for each.

THEORY

A passive *R-C* low-pass filter attenuates signals above the cutoff frequency and passes frequencies below the cutoff with little or no attenuation. A simple *R-C* filter is called a first-order filter and the attenuation above the cutoff frequency ($f_c = 1/(2\pi RC)$) is at the rate of -20 dB per decade. In passive filters the load resistance tends to increase the cutoff frequency and reduces the output voltage. An op-amp active filter eliminates the loading effect on the *R-C* circuit because of the very high input impedance and can provide voltage gain for the filter circuit.

PROCEDURE Part 1–Low Pass Filter

The active low-pass filter is shown in Figure 16.1 and the *Breadboard* model is shown in Figure 16.2. The capacitor C_1 (branch 8) is connected in parallel with the feedback resistor R_f (branch 7). At low frequencies, the capacitive reactance is very high and has little effect on the voltage gain of the circuit ($A_v = Z_f/R_1$). At high frequencies the reactance of the capacitor is low and reduces the output voltage gain of the circuit. Note that the compensating capacitor discussed in the previous exercise has been omitted, since it is very much smaller than the filter capacitor used here.

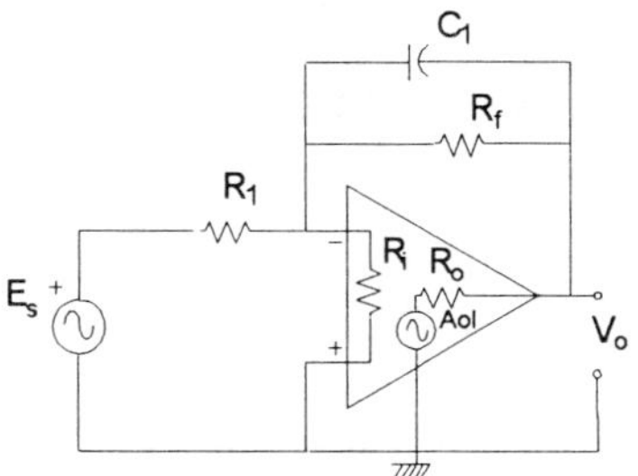

Figure 16.1

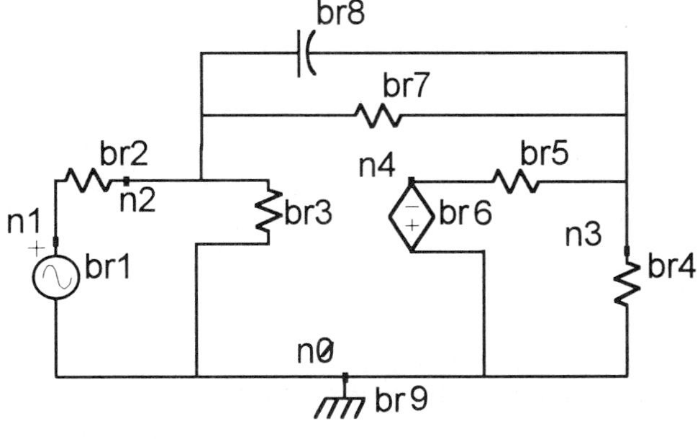

Figure 16.2

Net List for the Circuit

Br. 1: E = 10 mV/0°
Br. 2: R = 10 kΩ
Br. 3: R = 2 MΩ
Br. 4: R = 10 kΩ
Br. 5: R = 75 Ω
Br. 6: VCVS, A_v = 200,000
       control branch 3
Br. 7: R = 100 kΩ
Br. 8: C = 530 pF
Br. 9: voltage reference

175

Using *Breadboard* in the AC mode, draw the circuit model of the active low-pass filter shown in Figure 16.2. Obtain a Bode plot of the voltage output (V_{node3}) over a frequency range of 100 Hz to 100 kHz.

Using a straight-line approximation, obtain the cutoff frequency and the rate of roll-off from the graph.

The cutoff frequency (f_c) is _____________ Hz.

The cutoff frequency is calculated as $f_C = 1/(2\pi R_f C_1) =$ _________ Hz. ($R = 100$ kΩ, $C = 530$ pF)

The phase angle of the output voltage at the cutoff frequency is _______ °.

From the graph the roll-off is _____________ dB/decade.

Solve the circuit for a frequency in the pass band region of the filter (1 kHz).

The voltage gain of the amplifier is $A_v = (V_{node\ 3})/(V_{node\ 1}) =$ _____________.

By calculation the gain is $A_v = R_f/R_1 =$ _________________________.

PROCEDURE Part 2 – High Pass Filter

Repeat the procedure above for the high-pass active filter shown in Figure 16.3.

The gain for the circuit is $A_V = R_f/R_1$.

The cutoff frequency is $f_c = 1/(2\pi R_1 C_1)$.

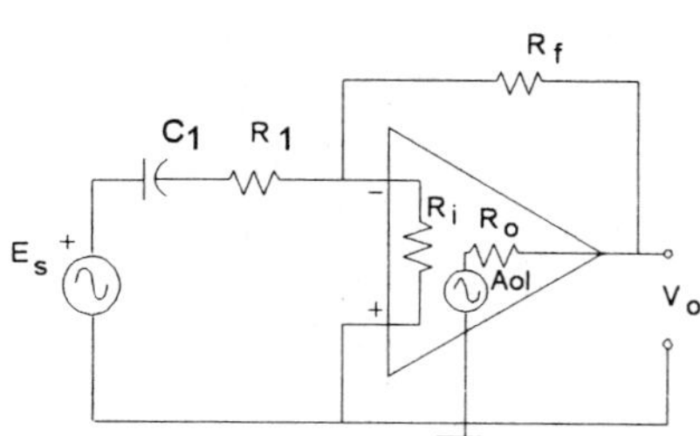

Figure 16.3

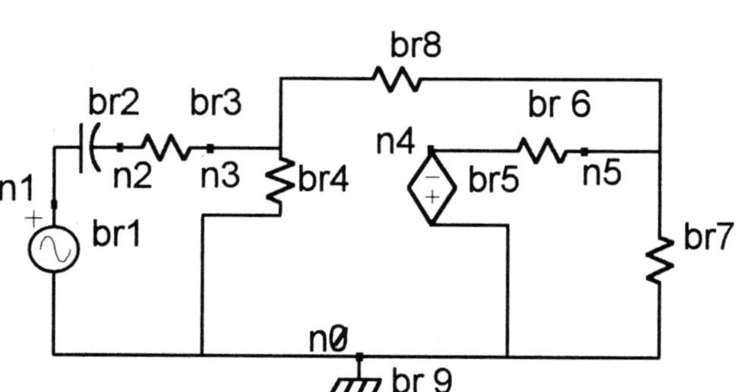

Figure 16.4

Net List for the Circuit

Br. 1: E = 10 mV/0°
Br. 2: C = 0.053 μF
Br. 3: R = 10 kΩ
Br. 4: R = 2 MΩ
Br. 5: VCVS, $A_v = 200,000$
         control branch 4
Br. 6: R = 75 Ω
Br. 7: R = 10 kΩ
Br. 8: R = 100 kΩ
Br. 9: voltage reference

Using *Breadboard* in the AC mode, draw the circuit model of the active low-pass filter shown in Figure 16.4. Obtain a Bode plot of the voltage output (V_{node5}) over a frequency range of 10 Hz to 10 kHz. Using straight-line approximation, obtain the cutoff frequency and the rate of roll-off from the graph.

The cutoff frequency (f_c) is ______________ Hz.

The calculated value is $f_c = 1/(2\pi RC) =$ ______________ Hz.
($R = 10$ kΩ, $C = 53$ nF)

From the graph the roll-off is ______________ dB/decade.

Solve the circuit for a frequency in the bandpass portion of the filter (2 kHz).

The voltage gain of the amplifier is $A_v = (V_{node\ 5})/(V_{node\ 1}) =$ ______________ .

By calculation the gain is $A_v = R_f/R_1 =$ ______________ .

PROCEDURE Part 3 – Second Order High-Pass Filter

A higher rate of roll-off (40 dB per decade) can be achieved by including R-C filter networks in both the input and feedback circuits of the amplifier as shown in Figure 16.5. The input R-C filter is made up of R_1 and C_2 and the feedback filter is R_2 and C_1. At the high frequencies both capacitors act as short circuits and the signal is passed to the output with practically no attenuation. To obtain a flat output (Butterworth) response the following relationships must be met.

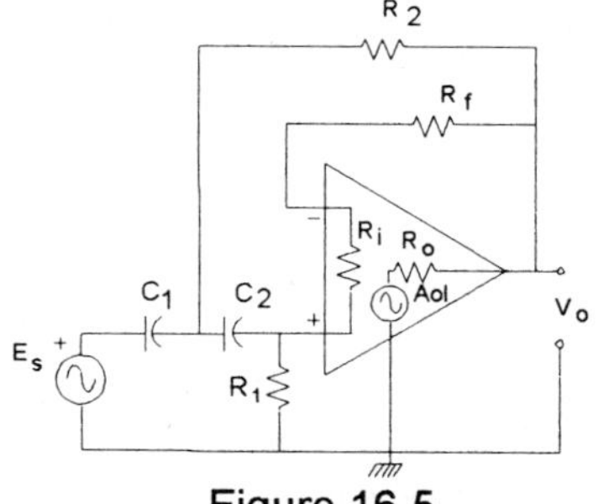

Figure 16.5

$$f_c = 1/(2\pi R_1 C_1) \qquad R_1 = R_2 \text{ and } R_f = 2R_1 \qquad C_2 = 2C_1$$

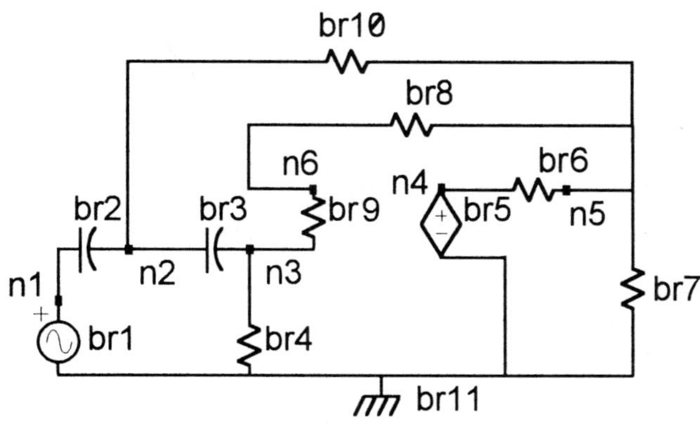

Figure 16.6

Net List for the Circuit

Br. 1: E = 10 mV/0°
Br. 2: C = 10 nF
Br. 3: C = 10 nF
Br. 4: R = 22516 Ω
Br. 5: VCVS, A_v = 200,000
      control branch 9
Br. 6: R = 75 Ω
Br. 7: R = 500 Ω
Br. 8: R = 22516 Ω
Br. 9: R = 10 MΩ
Br. 10: R = 11256 Ω
Br. 11: voltage reference

Using *Breadboard* in the AC mode, draw the circuit model of the active low-pass filter shown in Figure 16.6. Obtain a Bode plot of the voltage output (V_{node5}) over a frequency range of 100 Hz to 10 kHz. Using straight-line approximation, obtain the cutoff frequency and the rate of roll-off from the graph.

The cutoff frequency (f_c) is ______________ Hz.

The calculated value is $f_c = 1/(2\pi RC) = $ ____________ Hz.
($R = 22516\ \Omega$, $C = 10$ nF)

From the graph, the roll-off is ____________ dB/decade.

Solve the circuit for a frequency in the bandpass region of the filter (1 kHz).

The voltage gain of the amplifier is $A_v = (V_{node\ 5})/(V_{node\ 1}) = $ ____________ .

PROCEDURE Part 4 – Band Pass Filter

A band-pass filter is made up of a combination of the low-pass and high-pass filter circuits and is shown in Figure 16.5. Repeat the above procedure for this band-pass active filter.

The lower passband frequency $f_l = 1/(2\pi R_f C_1)$
The upper passband frequency is $f_h = 1/(2\pi R_2 C_2)$
The voltage gain $= A_v = R_f/R_2$

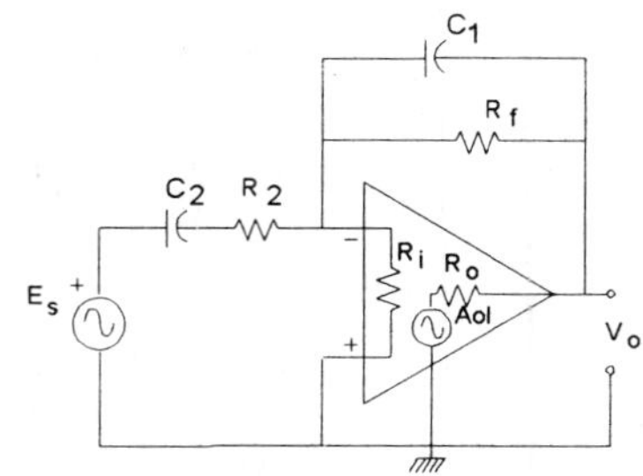

Figure 16.7

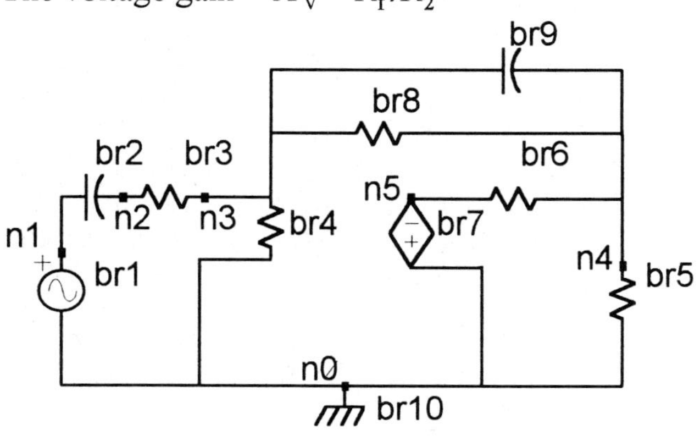

Figure 16.8

Net List for the Circuit

Br. 1: $E = 10$ mV/0°
Br. 2: $C = 530$ pF
Br. 3: $R = 10$ kΩ
Br. 4: $R = 2$ MΩ
Br. 5: $R = 33$ kΩ
Br. 6: $R = 75\ \Omega$
Br. 7: VCVS, $A_v = 200{,}000$
     control branch 4
Br. 8: $R = 100$ kΩ
Br. 9: $C = 53$ pF
Br. 10: voltage reference

Draw the circuit of Figure 16.8 and obtain a Bode plot of the voltage output (V_{node4}) over a frequency range of 1 kHz to 1 MHz. Using straight line approximation, obtain the bandwidth of the filter from the graph.

The lower and upper cutoff frequencies are $f_l = $ ________ Hz and $f_h = $ ________ Hz.

The bandwidth is $(f_h - f_l) = $ ____________ Hz.

From the graph, the roll-off is ____________ dB/decade.

The voltage gain of the amplifier at 30 kHz is $A_v = (V_{node\ 4})/(V_{node\ 1}) = $ ____ .

Second-Order Band-Pass Active Filter

OBJECTIVE

To obtain the frequency response curve for a second-order active band-pass circuit.

THEORY

To obtain better roll-off characteristics of filters, second-order filter circuits are employed that provide –40 dB/decade roll-off. These filters are classified as Chebyshev, Butterworth, or Bessel circuits depending on the flatness of the response curve and the sharpness of the cutoff.

To obtain a band-pass filter, two filter stages are used. The first stage is a high-pass filter and the second stage is a low-pass filter. Figure 17.1 shows the first stage with the filter capacitors in series; their value causes a cutoff frequency of 248 Hz. The second stage, with the capacitors in parallel, has a cutoff frequency of 2840 Hz. Filter theory requires that the gain of each stage be 1.586 to obtain the flatness of the Butterworth type. Using standard-value resistors, the gain of each stage is 1.57, providing an overall voltage gain of 3.16.

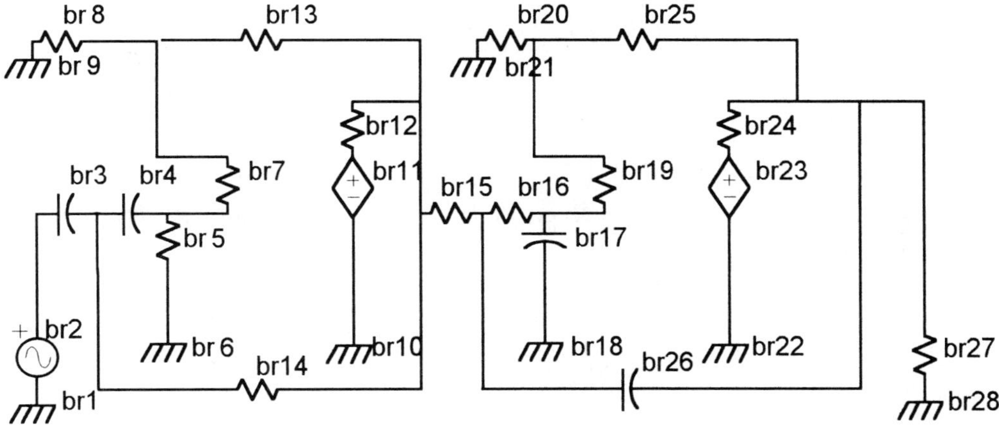

Figure 17.1

PROCEDURE

The model for the second-order band-pass filter is shown in Figure 17.1. Draw the circuit for the values given in the net list.

Net List for the Circuit

Br. 1: voltage reference
Br. 2: $E = 2$ V/0°
Br. 3: $C = 0.01$ µF
Br. 4: $C = 0.01$ µF
Br. 5: $R = 5.6$ kΩ
Br. 6: voltage reference
Br. 7: $R = 2$ MΩ
Br. 8: $R = 47$ kΩ
Br. 9: voltage reference
Br. 10: voltage reference
Br. 11: VCVS, $A_v = 200,000$
     control branch 7
Br. 12: $R = 75$ Ω
Br. 13: $R = 27$ kΩ
Br. 14: $R = 5.6$ kΩ

Br. 15: $R = 5.6$ kΩ
Br. 16: $R = 5.6$ kΩ
Br. 17: $C = 0.1$ µF
Br. 18: voltage reference
Br. 19: $R = 2$ MΩ
Br. 20: $R = 47$ kΩ
Br. 21: voltage reference
Br. 22: voltage reference
Br. 23: VCVS, $A_v = 200,000$
     control branch 19
Br. 24: $R = 75$ Ω
Br. 25: $R = 27$ kΩ
Br. 26: $C = 0.1$ µF
Br. 27: $R = 33$ kΩ
Br. 28: voltage reference

Solve the circuit for a frequency of 1 kHz.

Calculate the gain of the high-pass stage:

$$A_{vh} = \underline{\hspace{4cm}}$$

Calculate the gain of the low-pass stage:

$$A_{vl} = \underline{\hspace{4cm}}$$

Obtain a graph of the response curve ($V_{\text{node 11}}$) for a frequency range of 100 Hz to 10 kHz.

From the graph, determine the lower and upper cutoff frequencies.

$$f_l = \underline{\hspace{3cm}} \text{Hz} \qquad\qquad f_h = \underline{\hspace{3cm}} \text{Hz}$$

The bandwidth of the filter is:

$$BW = (f_h - f_l) = \underline{\hspace{3cm}} \text{Hz}$$

From the graph, calculate the dB roll-off of the filter.

$$\text{roll-off} = \underline{\hspace{4cm}} \text{dB/decade}$$

BREADBOARD REFERENCE

Preface

The *Breadboard* program was developed as a teaching/learning instrument for introductory courses in electrical engineering. The primary purpose was to enable students to easily obtain solutions to DC and AC circuits as encountered in undergraduate courses. The objective was to provide a computer analysis program that was very easy to learn and use so that the student would not be distracted from the study of circuit analysis by the complexity of the program. The program does not obviate the need to study the solution results of the circuit in order to develop the concepts and intuitive perception of circuit operation that a course in circuit analysis is intended to develop. It facilitates and provides the motivation for the achievement of that goal.

More advanced circuit-analysis programs are available on the market at the engineering design level. These, however, are usually at the graduate level, requiring a thorough knowledge of electrical engineering principles, the mastery of a bulky manual, and computer literacy. The *Breadboard* program can be taught to students within a single lecture period or utilized by students after a short demonstration. Emphasis can then be placed on the use of the solution results, rather than the program requirements, to obtain the solution. The scope and depth of the subject material can be extended and the teaching/learning of circuit behavior enhanced.

Drawing the circuit on the screen simulates the actual bread boarding of the circuit in the laboratory and provides a direct correspondence to the circuit diagram. The graphing capabilities of the program also permit the study of circuit solutions as a function of the variation of a circuit parameter. Analysis of this type in the laboratory would otherwise require a great deal of time. Laboratory experiment simulations can be effected prior to the event, and the experimental results can be compared to the program results.

The program is intended as an aid and an enhancement of presently used methods of teaching circuit analysis. It provides a mechanism for introducing computer-assisted learning into the courses and the potential for new educational approaches.

Getting Started

System requirements

> IBM PC or compatible computer
> 256K RAM
> 3.5" disk drive
> IBM Color Graphics Adapter (CGA) card or equivalent
> Printer (optional)

Running the program

> 1. Boot the computer with DOS 2.10 or higher version.
> 2. Insert the *Breadbord* program disk in drive A:.
> 3. Type BB and press return.

Adding the command line modifier **/B** will force the display to black and white for those systems having only black and white monitors.

Hard-disk users may wish to copy all the files on the program disk to a directory. From that directory, type BB.

The following options will appear on the screen.

```
┌─────────────────────┐
│     DC ANALYSIS     │        AC ANALYSIS                 EXIT
└─────────────────────┘
```

4. Use the left/right cursor-control keys to select either the DC or AC analysis option and press <ENTER>. The main menu options will appear on the top line of the screen as follows:

```
┌──────┐
│ DRAW │ SOLVE   SAVE   LOAD   NEW   PRINT   GRAPH   EXIT
└──────┘
```

Use the left or right cursor-control keys to select the desired option and then press <ENTER>.

The second line of the screen is the prompt line and will always indicate to the user the action or input required. The user should refer to the prompts until the user becomes familiar with the operation of the program. Each of the menu options is described in detail in the following sections.

The DRAW Option

The DRAW option enables "bread boarding" of the circuit on the screen. When this option is chosen, the screen is filled with a grid of dots, indicating the possible positioning of components.

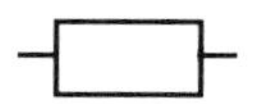

Component Cursor

A component cursor is located on the screen between two adjacent dots at a convenient starting position for drawing the circuit. It has not been placed in the lowest row of dots, since a chassis common symbol is required with every circuit and it is usual (although not necessary) to place the common at the bottom of a circuit diagram.

The component cursor can be moved to any vertical or horizontal location of the screen by using the left/right or up/down cursor-control keys. The movement of the cursor has been designed in a logical manner, preempting the desired action of the user. For the first-time user, it would be useful to move the cursor all around the screen to become familiar with its movements, especially when the limits of the screen are exceeded.

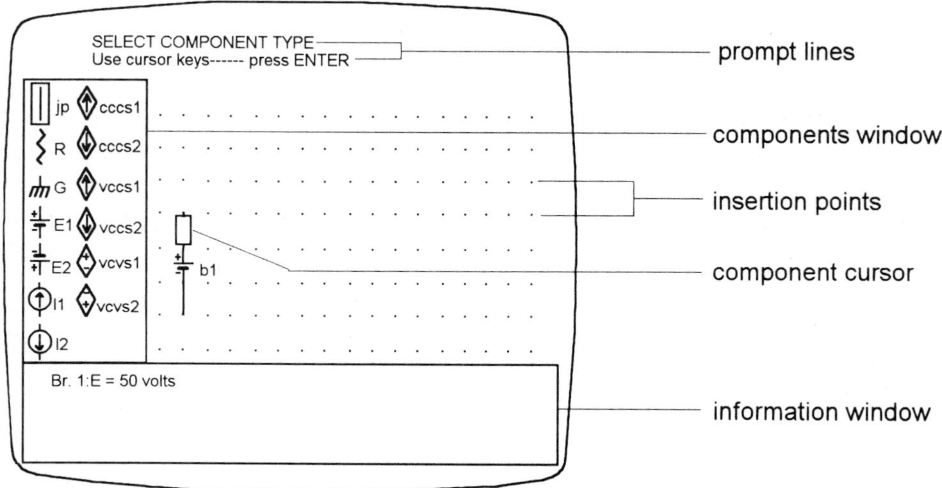

Although the components of the circuit may be drawn in any order, the preferred drawing sequence is to follow the circuit from source to load in the direction of conventional current flow in the circuit. To begin drawing, move the cursor to the desired starting location where a component is to be inserted. Press the <ENTER> key. A Component Window will appear to the left of the screen. Choose the desired component by using the cursor-control keys, and then press <ENTER>.

The chosen component will be drawn in the position of the drawing cursor. The prompt line at the top of the screen will request the numeric value of the component to be entered. Entering its value from the keyboard will cause a branch number to be assigned to the component on the screen. The component branch number, type, and value will be displayed in the information window as well.

The program will accept free-form entry of numeric values. For example, the number 10,500 may be entered as:

    10500   (numeric form)

    10.5E+3   (scientific notation)

    10.5k   (unit prefix form)

Other accepted unit prefixes are G, M, k, m, u, n, and p.

Do not enter the units for the component since the program recognizes the component that has been selected and takes the units by default. For example, to enter the value of 0.01 micro Farads for a capacitor, the user would enter 0.01u.

The program performs all calculations to double precision although the display is set to show only four significant figures. A value entered with a maximum 14 digits of precision is accepted as such even though the display shows only 4 digits.

Note: At least one chassis common symbol must be included on the diagram. This point on the circuit is used to establish the reference point to which all voltage measurements are taken.

Do not leave components unconnected on the diagram. To represent an **open circuit** use a very high value resistor equivalent to an open circuit, for example, 10^{12} Ω.

No **short circuits** are allowed. A short-circuit condition can be represented by a resistor whose value would be the resistance of an actual short, for example, 0.0001 Ω.

The Information Window at the bottom of the screen can display the value of 10 components at one time. If the window is full, further component entries will cause the information automatically to scroll upwards to display the latest entry. At any time during the DRAW option, the information in the window can be scrolled by pressing the cursor PgUp or PgDn keys.

When the drawing is completed press the <ESC> key to return to the Main Menu. Any of the menu options may be chosen at this point.

Editing the Schematic

If an error has been made in placing a component or entering its value, a correction is easily made. While in the DRAW mode, place the component cursor over the component to be modified and press <ENTER>. The component will be erased from the drawing and the information window will show that the component has been deleted. A new component may then be placed in that position and its new value entered.

The editing feature may also be used to erase complete branches of the circuit. Erase the jumpers as well as the components if they are part of the branch. Branches may also be added at any time using the DRAW option of the Main Menu.

The SOLVE Option

To obtain the solutions for the circuit, choose the SOLVE option of the Main Menu. (Use the cursor keys to select, then press <ENTER>.) The program may require several seconds to obtain the solution, as the circuit is first checked for continuity.

The results of the analysis will appear in the information window at the bottom of the screen. It may be necessary to scroll through the results, using the PgUp and PgDn cursor control keys to view them all. The following results of the analysis are available in the information window:

> Branch currents
> Branch power
> Node voltages (referenced to chassis common)
> Branch voltages (voltages across components)
> Volt-amperes reactive (for AC circuits)

To return to the Main Menu, press <ESC>.

Polarity Convention

In addition to the branch number being printed on the screen, additional information (node numbers) will also be placed on the schematic during analysis. The solution results provided by the program give the direction of the current flow (conventional current) and the voltage polarity across each component, as well as the magnitude (and phase angle for AC circuits).This result would be read as the current through the component of branch 3 is flowing from node 2 to node 3 and has a magnitude of 11.86 mA at a phase angle of 19.7°.

The direction of the current flow or voltage polarity is given in square brackets as the **node-from** to the **node-to**, as shown in the example below. This is equivalent to connecting a meter with the reference lead (common) to the node-to point.

Br. 3, [2–>3] 11.86 mA/19.7 deg.

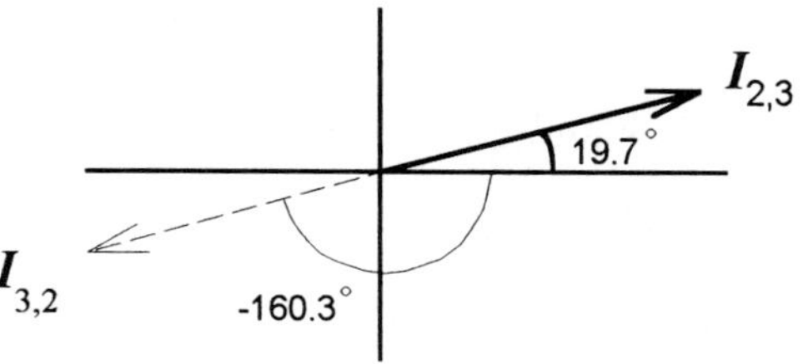

In some circuits, such as series R-L-C and multi-source AC circuits, the direction of the current or the voltage polarity provided by the program may appear to be reversed with respect to phase angle. In the following example, the phase angle is reversed by 180° but the direction of measurement as indicated by the node numbers gives this information. For the above example, the phase angle would be $(19.7°–180°) = –160.3°$. Both solution designations are correct.

Phasor Diagram

Br. 3, [3–>2] 11.86 mA/–160.3 deg.

For a voltage across a component, the result below would be read as the voltage across the component of branch 8 is 87.42 mV as measured from node 6 with respect to node 7. This notation is consistent with conventional double-subscript notation.

Br. 8, [6–>7] 87.42 mV

The SAVE and LOAD Options

The SAVE option will request a filename for the circuit. The filename may consist of up to eight alphanumeric characters. The program will automatically append the extension ".CKT" onto the filename. The circuit description will then be saved under the given filename in the current (default) directory in which *Breadboard* is running. A path specification may precede the filename if other than the default directory is to be used to store the circuit description.

The LOAD option is used to recall a circuit configuration from disk. The user is prompted to specify a directory path or if <ENTER> is pressed the default path will be used. The program will automatically list the filenames with the extension .CKT in the chosen directory. The user then selects the filename of the circuit to load by using the cursor keys to select and then pressing <ENTER>. The program will draw the circuit on the screen and display the description of the components in the information window. The user may then select any of the options from the Main Menu.

The NEW Option

Selecting the NEW option will erase an existing drawing from the screen and erase the information window, allowing the user to start fresh. Since all the information pertaining to the circuit is erased from memory, the user should save the circuit before selecting the NEW option if the circuit is to be recalled for future reference.

The PRINT Option

The PRINT option provides the capability of obtaining hard copy of a number of selections which appear in a selection window when the PRINT option is chosen.

```
   Schematic

   Net List

   Br. Currents

   Br. Power

   Node Voltage

   Br. Voltage

   ** PRINT **

   <ESC> to menu
```

The Print Menu

To choose one or more items to be printed, move the blinking selection box to the desired item and press <ENTER>. A non-blinking box will remain on that item to indicate that it has been chosen. If more than one item is desired, move the blinking box to the next item and again press <ENTER>. When all the desired items have been selected, move the blinking box to the **PRINT** and press <ENTER>. The selected items will be printed.

If an item has been chosen in error, the selection may be canceled by moving the blinking box back to that item and pressing <ENTER> prior to selecting the **PRINT**. An alternative would be to press <ESC> to return to the Main Menu and repeat the process.

If the circuit has not been solved and a printout of the solution values is selected, the program will obtain the solutions first and then print out the selected results. Thus it is possible to draw the circuit and go directly to the print option to obtain a printout of the solutions.

The CALC Option

In AC mode an additional option is CALC. Circuit parameters are often expressed in terms of reactance or impedance whereas *Breadboard* requires the values to be entered as resistance, inductance, or capacitance. This options provides the following conversions to be made .

$$L \text{ given } X_L$$
$$C \text{ given } X_C$$
$$R_S \text{ \& } C_S/L_S \text{ given } Z$$
$$R_P \text{ \& } C_P/L_P \text{ given } Y$$

The GRAPH Option

The GRAPH option will draw a screen graph of the branch current or voltage, or the node voltage or branch power as a function of the variation of a branch resistance, a voltage source, or current source and, in AC mode, the variation of the frequency of the source.

When this option is chosen, a representation of a graph appears in the information window to facilitate the choice of graph options available. The user will be prompted to provide information regarding the choice of the independent (X-axis) variable: resistance, voltage source, or current source. In the AC mode, frequency may also be selected for the independent variable. The selection is made using the up/down cursor keys to select the option and pressing the <ENTER> key. The user will be prompted at the top of the screen for the branch number of the variable and the range of variation.

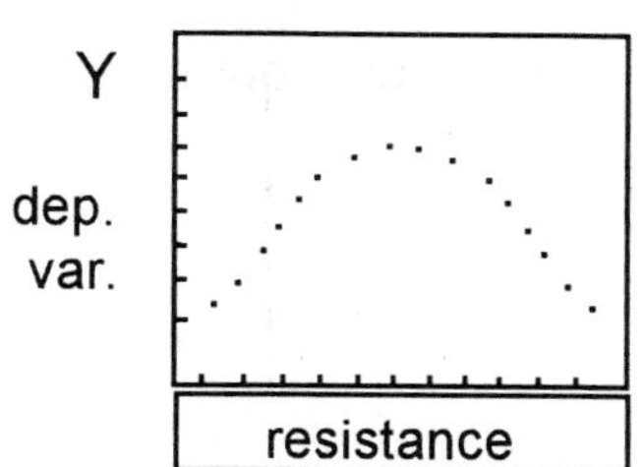

The Graph Option

The selection of the dependent variable (Y-axis) to be plotted is similarly chosen from the graph representation by using the up/down cursor keys followed by <ENTER>. The available options are branch current, branch voltage, node voltage, or power. The prompt line will ask for the branch number of the component whose parameter is to be plotted

In the AC mode, a Bode plot of dB versus frequency is also available as an option. The user is prompted to provide the voltage reference value required for the calculation of the dB value of gain or attenuation.

$$dB \;=\; 20\log\frac{V_{out}}{V_{ref}}$$

The program then performs all the necessary functions of scaling, solving the circuit for 120 points within the range of variation selected and plotting the graph on the screen. If the range of variation selected is greater than one decade, the program produces a semilogarithmic graph of from one to five cycles maximum as required.

On completion of the graph, the program will request whether a hard copy of the graph is desired or not. For a response of Y (yes), the program will ask whether the printer is an IBM, HP Laser, or EPSON type. The program uses specially designed characters in order to duplicate the screen graph. If the printer is not capable of accepting download characters, the graph can be printed out in text-mode graphics that does not, however, provide as good definition. Check your printer manual to determine if the printer has character download-enable capabilities, and check that the switch settings have been properly set. An alternative for obtaining hard copy of the screen graph would be to use the PrtSc function on the keyboard to obtain a screen dump of the graph. The DOS utility GRAPHICS.COM must be loaded prior to running *Breadboard.*

A tabular listing of the solution values (120 points) used to plot the screen graph can be obtained on the printer. This is useful if the degree of accuracy desired cannot be obtained from the screen graph. The user is prompted to respond Y (yes) or N (no) to print out the values. The printed results are single precision although the calculation of the results were performed as double precision.

Using Controlled (Dependent) Sources

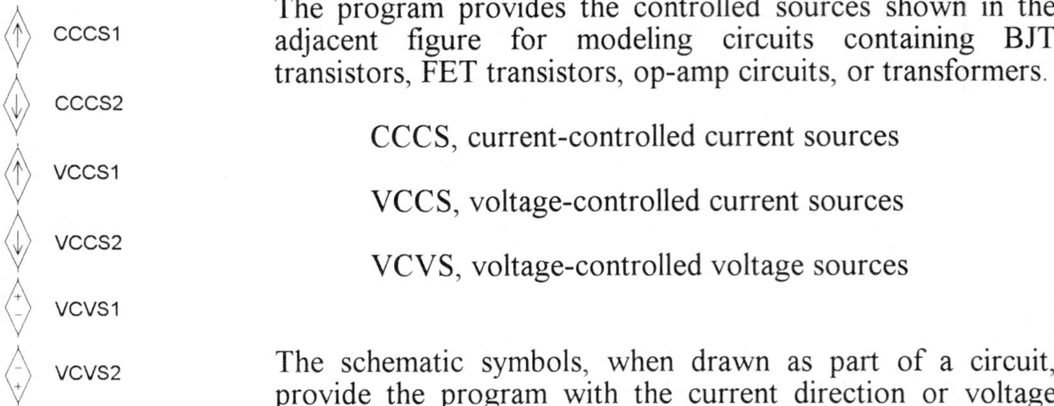

The program provides the controlled sources shown in the adjacent figure for modeling circuits containing BJT transistors, FET transistors, op-amp circuits, or transformers.

CCCS, current-controlled current sources

VCCS, voltage-controlled current sources

VCVS, voltage-controlled voltage sources

The schematic symbols, when drawn as part of a circuit, provide the program with the current direction or voltage polarity of these sources (conventional current flow). The polarity or current direction of the **controlling-branch** is also a determining factor for the controlled source. It is important to use the following convention to obtain the expected results of the circuit analysis.

The movement of the drawing cursor to the position where a **controlling-branch** is to be placed must be in the direction of the conventional current flow in that branch. In the example of Figure A1, the order of drawing the components in the control branch (branch 3) mesh must be clockwise, following the current flow.

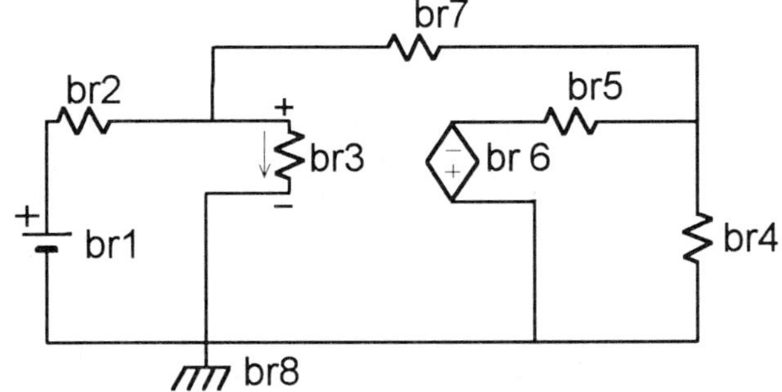

Note: If the above sequence is not followed, the program may reverse the voltage polarity or current direction of the controlled source and cause confusing or erroneous results in the circuit analysis.

Figure A1

The polarity or current direction for the **controlling-branch** indicated in Figure A1 is taken by the program to be the direction in which the branch was drawn. The polarities or current directions of the **controlled source** will then be as depicted by the schematic symbol selected for the source. A simple rule to follow is to draw the components in the order and direction of conventional current flow through them.